LTE Services

*Series Editor*
*Pierre-Noël Favennec*

# LTE Services

Jean-Gabriel Remy
Charlotte Letamendia

iSTE

WILEY

First published 2014 in Great Britain and the United States by ISTE Ltd and John Wiley & Sons, Inc.



ISTE Ltd
27-37 St George's Road
London SW19 4EU
UK

www.iste.co.uk

John Wiley & Sons, Inc.
111 River Street
Hoboken, NJ 07030
USA

www.wiley.com



Library of Congress Control Number: 2014947876

British Library Cataloguing-in-Publication Data
A CIP record for this book is available from the British Library
ISBN 978-1-84821-787-4

# Contents

# List of Figures

**CHAPTER 2**

# List of Tables

# Introduction

Long Term Evolution (LTE) is commonly marketed as fourth generation (4G). LTE and LTE Advanced have been recognized by International Telecommunications Union – Radiocommunications (ITU-R) and International Telecommunications Union – Telecommunications (ITU-T) as the principal solution for the future mobile communication networks standards. Thus, they are the framework of what marketing calls 4G and maybe also fifth generation (5G). They have registered logos:

**Figure I.1.** *LTE and LTE Advanced logo*

It seems interesting to look at the evolution of mobile communication systems from their appearance upto LTE. This move has obviously been driven by commercial motivations as well as by the extraordinary improvement of microelectronics, especially from the 1960s to the present day. Functionalities, computing power and miniaturization have drastically progressed, while cost has constantly decreased.

## I.1. Mobile communication systems: 0G, 1G, 2G, 3G, 4G and 5G

In this short introduction, many mobile communication systems will be omitted:

– military communications and public utilities communications;

– maritime and aviation communications;

– trunk systems and more generally all kinds of professional dedicated radio systems.

It does not mean that LTE will not have specific adaptations in order to fit the special requirements of such systems, especially for its radio interface, avoiding expensive developments being invested for a limited population of users.

Only public land mobile network (PLMN) will be considered: the so-called "4G" belongs to this category as long as LTE is used for public communication.

Also, the impressive list of various systems, which did not reach a high level of success, especially outside their country of origin, has been avoided.

The classification of mobile systems into generations is not strictly related to any given metrics or parameters. It corresponds to marketing considerations. Therefore, it is commonly agreed upon, both by industry and by academia, and hence conceived to be an unwritten standard.

### I.1.1. *Rationale*

Mobile communications have always been a wish for most of the people. Of course, at the beginning, the mobile networks have been invested for precise applications, such as military communications or professional management. The introduction of PLMN came later. But the requirements for mobile services are most common for public systems and more specific networks.

For a network addressing all citizens, the investment is very high, especially in research and development – millions of coded instructions have to be written and validated. Also, the precise areas where the service will be necessary have to be determined. Therefore, it is necessary to analyze what the customers are ready to pay for to avoid vain efforts and investments. Excluding applications that are just using the mobile network as a support, mobile services can be classified into three categories:

– Mobile telephony: the mobile subscriber wants to discuss in real-time with distant interlocutors, who are connected with either a fixed telephone or a mobile set. Telephony offers the possibility to get immediate up-to-date information as well as the means to discuss any difficult item. Up until now it has been the most "money making" application.

– Paging: by some means of collection of the information, the network offers the capacity to alert the mobile subscriber that something of interest is happening. The paging can be limited to a very simple binary signal – some tone or light – and the customer has to call an information center to get the message. It can also be accompanied by a short message, either written or vocal, giving the main details of the message. This paging is very popular and is now offered by the short message service (SMS) of Global System for Mobile communications (GSM) and further technologies. The SMS service is a "teleservice", which means that the operator must carry it to destination. The multimedia messaging service (MMS) delivers much richer information, but it is not as reliable, because the delivery of messages is not guaranteed by the network operator; it is supported by a "bearer service", the quality of service (QoS) is limited to the operator's commitment.

– The Internet, fax or any written dialog: in the latter case, the mobile network offers the possibility to carry the office environment of its customer anywhere. Like MMS, the Internet and Internet-like services are generally bearer services, which are sold with a certain grade of QoS.

For these services, the mobile network can provide two kinds of access:

– nomadic access: the service is available anywhere inside the coverage of the network, but the customer must be static or is allowed to move very little;

– full mobile access: the service is available when the customer is moving, eventually at any speed, again within the limits of the geographical coverage service.

The paradigm of mobile communications is simple to summarize:

– be able to be connected to and receive information from any calling party;

– be able to be connected to any called party;

– full bidirectional access and real-time exchange of information;

– be accessed anywhere, outdoor, indoor, in urban and rural environment;

– full bidirectional access at anytime.

Going into detail shows a lot of issues:

– size of the mobile device: devices such as smartphones or tablets such have limited space to support the broadband module; these days, the terminal can also be some communication part of a machine for machine to machine (M2M) communications;

– nature and content of information to be transmitted, i.e. full telephony, television or data transmission, bilateral or unilateral.

### I.1.2. *Short history of mobile communications, milestones*

#### I.1.2.1. *0G*

The systems that allow customers to communicate on the move depend on electronics and microelectronics technology. Therefore, before the mass production of semiconductors, only experimental services have been deployed. The first network appeared in the United States in 1940, with mobiles using electronic tubes for car mounted terminals. Connection to the called party was made by human operators, in a way similar to that ensured for maritime communications.

Between 1960 and 1980, quite a few mobile communication systems were designed and deployed for either telephony or paging. Most of the advanced countries installed a home-made network. These systems offered automatic dialing with a good communication quality, obtained with a frequency/phase modulation radio access network. The radio path consisted of narrow frequency channels – 30 kHz in Northern America and 25 kHz everywhere else in the world. With the advent of transistors, a few handheld mobiles were available, especially for paging.

Of course, the service was only operated by incumbent fixed telecommunication operators, which found a new service for wealthy customers.

*These systems will be called 0G.*

#### I.1.2.2. *1G*

During the 70s, some important innovations have brought a kind of revolution in the mobile communication world:

– computer driven frequency tuning (frequency synthesis) allowing us to reach with precision a given radio frequency channel among many with only one quartz oscillator. This technology opened the way to high-capacity systems in so-called analog technology – where each individual communication is allocated one (time division multiplex (TDM) or simplex) or two (frequency division duplex (FDD) or duplex) precise narrow band frequency channels – managing hundreds of radio frequencies instead of a few tens in the previous systems. With such a number of channels, the radio communication system becomes able to cope with a large number of customers. Also, frequency synthetization opened a way for massive production of handheld terminals:

– standardization and generalization of Signaling System No. 7 (SS7) designed for telephony, mainly the international version of ISDN;

– availability of microcomputers and computing chips offering greater speed and power for real-time processing, thus allowing us to implement sophisticated encoding, error correction and new transmission standards.

All these innovations were applied to new designs including some important breakthroughs:

– localization of the mobile terminal, which could be done manually, and automatically realized, in order to have the ability to route incoming calls;

– detection of the need for changing the communication in progress from one radio base station (one "cell") to another due to degradation of the radio link quality, and execution of the "handover" (US: hand off) to the other base station/cell, which is selected to provide a good quality communication.

With all these new developments, the cost of R&D skyrocketed and only a few systems could be studied and deployed with a worldwide impact. Among them two standards will dominate the market:

– First, the advanced mobile phone system (AMPS), designed by the Bell Labs with a prototype rollout installed in Chicago in 1978, serving more than one thousand customers. AMPS has been the first system to offer real-time seamless handover. This network probably shows the best possible design for a system where each individual communication carried by an individual duplex frequency modulation (FM) (or phase modulation (PM)) channel, each channel being given a narrow frequency bandwidth. The main features were standardized by the American National Standard Institute (ANSI). This AMPS system has the particularity of being able to modify channel spacing and FM excursion very simply, which allowed us to adapt it to various frequency configurations (channel spacing of 30 kHz in the USA and 25 kHz in Europe and Japan). This is achieved simply by modifying the clock frequency driving the network. In North America, it was the genuine AMPS (initially, A stood for American).

In Europe and Japan, it was a modified version with a 25 kHz channel spacing, called Total Access Communication System (TACS), Europe TACS (ETACS) and Japan TACS (JTACS)). Due to some specific US political process aiming at introducing competition, AMPS and TACS massive deployment was delayed to 1985.

– However, the Scandinavian countries joined their strengths and developed the Nordic Mobile Telephone (NMT) system. This standard is by far simpler than the AMPS/TACS in all aspects of the technology. The spread of NMT is somehow due to the above-mentioned American political process, which delayed the mass deployment of AMPS. NMT became available around 1982 and was immediately rolled out in all Scandinavian countries.

Nevertheless, due to its transnational origin, NMT introduced a very interesting feature: automatic international roaming.

Another cellular system of the first generation was designed and deployed in Germany (C-Netz) and France (Radiocom, 2000) and counted a few hundred thousand subscribers. There was also a Japanese home-made "cellular" system.

*These systems and their unlucky competitors are considered to be 1G.*

#### I.1.2.3. *2G*

In the 1980s, with the spectacular increase of the computing power of integrated circuits, technology continued to progress with many breakthroughs:

– Development of vocoders. In concordance with the design of very powerful processors. Instead of needing a bitrate of 64 kbps to correctly digitalize narrow band voice telephony as calculated from the ordinary Shannon sampling, a telephony 4 kHz analog signal can be coded with a very good quality with 12 kbps, and even 6 kbps

(GSM). For professional systems, vocoders provide a clear voice communication with a few hundred kilobits per second.

– Vocoders are the key to switch from analog FM (or PM) radio to full digital transmission for telephony. The compression of the voice signal is a question of processing power. Today, a very high quality sound can be coded with less than 10 kbps; and correct voice communications are now available for professional and military communications with a bitrate of less than 1 kbps.

– Development of identity chips. The 1G German C-Netz had introduced a device to dissociate the subscription from the mobile terminal hardware. Such chips make it possible to encrypt communications and protect customers' privacy. AMPS or NMT were identifying the mobile terminal by a number which was included inside it and was very easy to copy or modify; so, customers were often suffering from pirated use of their identity. Concerning the privacy of communications, 1G networks did not provide protection against eavesdropping.

In the meantime, continental European countries have been conscious of their technological backwardness compared with AMPS. In 1982 the "GSM" was created (at the beginning it was a "special mobile group" led by German FTZ and French Centre national d'études des télécommunications (CNET)), which was commissioned to study a revolutionary mobile system based on a fully digital radio access subsystem, since it was considered difficult to surpass AMPS as an analog system. This new system, also called GSM, passed through a lot of studies until 1991. Code division multiple access (CDMA), which was in the 1980s a spread spectrum technique in use for military purposes, was experienced in 1985. At that time, CDMA showed need for too much computing power, far over the performance of the available chips, thus a simpler process, time division multiple access (TDMA), was chosen.

In 1987, all countries of the European Union signed a Memorandum of Understanding (MoU), which was accepted afterward by all GSM operators, always labeled as MoU. In this MoU, these countries decided:

– to roll out a GSM coverage from 1991 onward using the common frequency bands which had been decided in 1979 for a common mobile system;

– to authorize without restriction automatic international roaming for GSM mobiles, all expenses being paid by the home country of the subscription.

GSM takes up the C-Netz innovation of selling the mobile terminal and the operator subscription separately, the latter being materialized by a SIM card, which is inserted into the mobile set. The chip of the SIM card controls all the telecommunication functions of the mobile and masters the encryption of the radio path for the calls.

GSM introduces a kind of paging with the "SMS", which became a very important part of the communications.

As a response to the introduction of GSM, the AMPS industry designed the D-AMPS (IS-136 standard), where AMPS channels are used in TDMA mode in order to increase the overall network capacity.

Beside the TDMA systems, the American society Qualcomm introduced its proprietary design based on a CDMA encoding, later called CDMA 2000, which was standardized as IS-95 by ANSI. This standard was adopted by South Korea, which had to solve a lot of difficulties.

And again, Japanese NTT developed and rolled out a TDMA system, called PDC. They also rolled out a simpler system called PHS, which is probably the first implementation of a multiple input multiple output (MIMO) antenna system.

*All these systems can be considered to be the 2G mobile standards.*

#### I.1.2.4. *3G, the need for fast data transmission*

Of course, as time passed, the technology of chips continued to improve drastically. During the 1990s it finally delivered processors having a sufficient computing power to cope with the Qualcomm CDMA mobile system.

In the 1990s, while GSM was being implemented all over the world including Northen America, the operators of fixed communications introduced the Internet services. At the beginning the available bitrate was limited to 50 kbps. Later it was increased to 10 Mbps downlink particularly with an Asymmetric Digital Subscriber Line (ADSL), provided the customer's home is located a few hundred meters from the central office. The industry of mobile communications decided to adopt the internet service in their strategy, even when the response from the subscribers' base surveys showed very little interest in telephony and SMS. GSM developed a "wart", called General Packet Radio Service (GPRS), supporting data transmission upto 50 kbps. In response, CDMA 2000 introduced data transmission upto 144 kbps. As an answer, GSM standardized Enhanced Data Rates for GSM Evolution (EDGE), providing upto 240 kbps, which was rolled out massively by ATT Wireless in the USA, where it was facing the competition of Verizon Wireless, the CDMA 2000 champion.

The way Qualcomm system manages data transmission makes it easy to reach good performances since the data flow and the telephony are transmitted by different networks, at least in the Evolution Data Optimized (EVDO) version. This conception answers the difficult challenge of mobility:

– telephony is a real-time communication, but accepts very short cuts, e.g. 300 ms; this is managed by a smooth handover process;

– data transmission in Transmission Control Protocol-Internet Protocol (TCP-IP) shows very poor performance if the flow is cut, as is the case when the mobile travels from one cell to another. In that case, a reselection is necessary and the usable bitrate is very poor

Considering that in a town like Paris the mobile terminals process an average of four handovers for a 2 min call, the network operator has to make a critical choice concerning the parameters of its network:

– either the parameter set favors telephony with a change of cell achieved as soon as possible to give the customer a very good telephony quality;

– or the parameters are stiffened and the mobile will drag its radio channel as far as possible in order to avoid reselection. It results in damaging the frequency planning, as well as creating poor quality telephone calls.

Of course, most of the GSM operators chose to favor telephone calls.

To examine what could be the future of mobile communications after the worldwide success of GSM, the European Union launched a consultation on the possible technologies which could be developed. Scandinavia pushed a variant of Qualcomm CDMA technology called wide band CDMA (WCDMA) very hard, which won the competition. This WCDMA technology immediately faced the issue of patents, since CEO of Qualcomm, who was a highly respected former professor of signal theory at MIT, had patented all possible implementation of CDMA. It also faced plenty of issues with the management of power, with the mobile needing too much energy, far more than GSM.

Nevertheless, the industry worked very hard and some 10 years later, beginning of the 2000s, the WCDMA, renamed High Speed Packet Access (HSPA) and HSPA+, could service data users correctly.

In the meantime, ATT had pushed in the 3rd Generation Partnership Project (3GPP) standard body, a variant of GSM, called EDGE, which had been rolled out by all GSM operators. The advantage of EDGE for the network operator is to keep the base stations of GSM for coverage and reuse the same backhaul infrastructure instead of deploying a new network. EDGE, described above, is a modification of GPRS (changing the modulation on the radio path) and provides 200 kbps and more.

*EVDO, WCDMA and EDGE could be considered as the 3G mobile systems.*

I.1.2.5. *4G*

As seen above, the work on *Universal Mobile Telecommunications Service* (UMTS) finally produced a competitive system, called HSPA, then HSPA+, that reached upto 7.2 Mbps, and even 14.4 Mbps per cell.

In the 3GPP studies, besides promoting EDGE, ATT called for a completely new system, strictly dedicated to mobile data transmission. Their concept at the beginning was to design something completely new with no backward compatibility with previous systems. The new system would be completely based on IP and would adopt a simple architecture. This project was called "LTE" and was the answer to ITU request of a future mobile system (called FPLMNTS in the 1990s, denomination replaced by IMT2000, then IMT Advanced).

The LTE standard was finalized only in 2008 with the release 8 of 3GPP.

When definitively designed in a viable release, LTE was immediately adopted by Qualcomm CDMA followers, especially Verizon, which will abandon CDMA 2000 progressively. So, *de facto,* LTE became the only standard of mobile communications for the future. The system is now widely deployed, mainly in Northern America with over 100 million subscribers there, and represents a very strong industry.

Having been badly fleeced with intellectual property rights (IPR) in the UMTS case by Qualcomm, and less seriously by Motorola for GSM, 3GPP's "individual members" exert a certain control on the ETSI IPR database. In 2012, 50 companies had declared holding essential patents covering some parts of the LTE standards. Nevertheless, these declarations are left to the goodwill of the companies, even if at each TSG meeting participants are invited to declare their patents with a certain solemnity.

*The 4G mobile system follows the LTE standard.*

I.1.2.6. *5G*

What about the "5G" on which many publications have already been edited? LTE radio access subsystem is based on different avatars of Orthogonal frequency-division multiplexing (OFDM), technology described in 1982 by the CCETT laboratory of Rennes (France). OFDM is now recognized as the best technique for transmitting high bitrate flows of data on wideband radio channels. It has been adopted for the last versions of Wi-Fi (IEEE 802.11n and further), by WiMAX (IEEE 802.16, beyond "e"), Communications over Power Lines (CPL, in UK power line communications (PLC),) and television broadcasters with the DVB-S2 and DVB-T2.

5G is probably what is considered as IMT-Advanced with the following requirements:

| Item | IMT-Advanced Requirement | LTE-Advanced Projected Capability |
|---|---|---|
| Peak Data Rate Downlink | | 1 Gbps |
| Peak Data Rate Uplink | | 500 Mbps |
| Spectrum Allocation | Up to 40 MHz | Up to 100 MHz |
| Latency User Plane | 10 msec | 10 msec |
| Latency Control Plane | 100 msec | 50 msec |
| Peak Spectral Efficiency DL[133] | 15 bps/Hz | 30 bps/Hz |
| Peak Spectral Efficiency UL | 6.75 bps/Hz | 15 bps/Hz |
| Average Spectral Efficiency DL | 2.2 bps/Hz | 2.6 bps/Hz |
| Average Spectral Efficiency UL | 1.4 bps/Hz | 2.0 bps/Hz |
| Cell-Edge Spectral Efficiency DL | 0.06 bps/Hz | 0.09 bps/Hz |
| Cell-Edge Spectral Efficiency UL | 0.03 bps/Hz | 0.07 bps/Hz |

**Table I.1.** *Mobile broadband explosion*

*(Source: mobile broadband explosion: the 3GPP wireless evolution, Rysavy Research/4G Americas, August 2012)*

Assuming that the "5G" will be allocated a large amount of spectrum (e.g. more than 20 MHz, or upto 100 MHz if such a quantity of spectrum can be found), the radio transmission scheme could be improved or upgraded as has been the case for the change from Digital

Video Broadcasting Terrestre (DVB-T) to DVB-T2 and from Digital Video Broadcasting Satellite (DVB-S) to DVB-S2. From the measured performance of DVB-T2 an overall bitrate of 100 Mbps available for the individual subscriber could be expected with a reasonable spectrum allowance.

1 Gbps would probably need a big part of spectrum, which could not be foreseen some 10–20 years ago, except if the system adopts frequencies above 3 GHz and restricts mobility.

The difficulty to make a valuable forecast comes from 2 sides:

– most smartphones and also mobiles can also communicate through Wi-Fi, and this communication cost nothing to the subscriber nor to the operator. This will probably impact the business plan of a possible 5G;

– the development cost of such systems reaches very high levels, only very few industrial companies can finance the necessary R&D. To date, only two or three companies are competing for delivering the LTE infrastructure.

LTE Advanced has been accepted as IMT-Advanced relevant solution in November 2010. LTE_advanced must be both backward and forward compatible with existing LTE. Devices must operate on both kinds of networks.

A few operators and manufacturers claim that their research and development laboratories have already tested IMT-Advanced solutions with:

– wider bandwidth support for up to 100 MHz via aggregation of 20 MHz blocks (carrier aggregation);

– uplink MIMO (two or four transmit antennas in the device);

– higher order downlink MIMO of up to 8 by 8 as described in release 10;

– coordinated multipoint transmission (CoMP) with two proposed approaches: coordinated scheduling and/or beamforming, and joint processing/transmission (in release 11);

– heterogeneous network (Het-net) support including enhanced inter-cell interference coordination (eICIC);

– relay.

Figure I.2 shows the evolution flow:

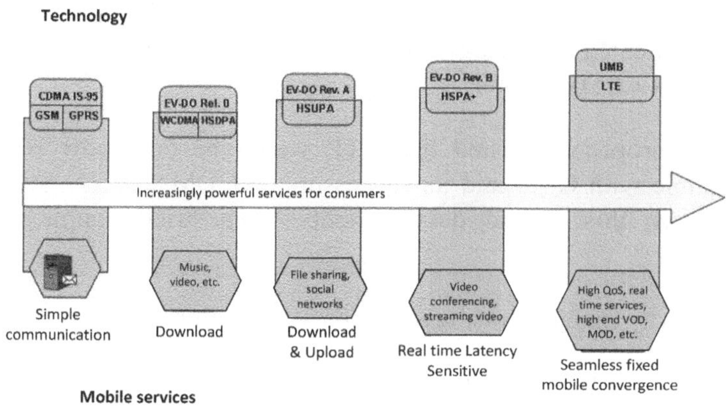


**Figure I.2.** *The LTE project: milestones. Short history of the birth of a worldwide standard*

*What is now called LTE had been proposed in 1998 as a successor to GSM, but was not chosen and 3G has been based on WCDMA mainly.*

LTE has been developed by 3GPP.

**Figure I.3.** *3GGP logo*

After a long and difficult process in the 3GPP, ATT engineers succeeded to introduce LTE as a work item (3GPP, http://www.3gpp.org/specifications). Their concept was to describe a "green field" system, which would have replaced all existing techniques and would provide, at last, a worldwide accepted technology. The emergence of LTE has been delayed by European actors, both mobile operators and industrial manufacturers, which had spent a huge amount of money for WCDMA, the 3G system called UMTS. Operators had to pay enormous fees for UMTS licenses; industrial companies had to pay high patent dues to Qualcomm for the use of a patented technology, even if UMTS is quite different from the Qualcomm's CDMA2000.

The Europeans insisted that LTE would be (and now is) quite compatible with GSM and its successors (WCDMA or TD-SCDMA, even when this second development seems strictly applicable to China).

LTE is by many sides a revolutionary technology.

Parallel to the 3GPP work, ITU-T set a work item for the future mobile communication system, first called FPLMNTS then renamed IMT to finish with IMT2000, followed by IMT Advanced.

LTE release 8 is the first standard describing a working technology. Issued in 2008, this release 8 showed a system, which had no telephony service and was fully dedicated to Internet communications, and therefore had to fall back to GSM or WCDMA for telephony if not leaving the task to OTT applications. LTE was and is a pure Internet-based system deliberately designed for packet data communications. Packet communications are no longer a kind of wart added to a telephony system, like GPRS or EDGE for GSM, but the principal objective of a full "Internet multimedia system".

LTE had to wait for release 11 (at the end of 2012) to be able to provide a telephony service. Nevertheless, it has been recognized as the practical incarnation of IMT Advanced in 2010. This recognition has been eased by the renunciation of Qualcomm's 3GPP2, the experts of which could not follow the breakthroughs obtained by the hundreds (maybe thousands) of engineers working on LTE. Moreover, the champion of CDMA2000, the American operator Verizon Wireless, was among the first in the world to roll out LTE.

Some features will only be available in release 12 (at end of 2014) and probably later. It is expected that the "change requests" on LTE standards will continue to flourish until 2020.

But now, the only competing standard is WIMAX, the IEEE 802.16 standard, which has evolved recently to somehow adopt the same technological choices as LTE on the radio path, especially OFDMA. Also, Wi-Fi, 802.11, in its last avatar has also switched to OFDMA. Wi-Fi is more in a position to compete since it has not at all the same business model, offering mainly free communications carried by unlicensed frequencies. The advantage of LTE on all competitors is that it is the only system which has a fully described and standardized core network, based on IMS.

LTE has been the substrate of the frequency battle in ITU-R world radio conference 2007 with the American pushing for allocating the 700 MHz band to mobile communications (i.e. LTE) and the European deciding to offer to LTE high frequencies such as 2.6 GHz, 3.8 GHz and even higher. These frequencies may only be suitable for "Wi-Fi like" communications because at these high frequencies tens of thousands of base stations are needed with little chance to cover each more than one stretch of a street. They are obviously inadequate for the coverage of wide spaces, like a full country. Of course, on the opposite, the 700 MHz is excellent for the coverage of wide areas, e.g. the Middle West area. In urban areas, frequencies under 1 GHz are also much more efficient, as they better penetrate the buildings or the underground.

The consequence of these choices is that LTE/4G is, in 2014, mainly rolled out in the United States and Canada using 700 MHz and 1800 MHz base stations. The market of many tens of millions of subscribers is a strong incentive to provide cheap and excellent smartphones following the American choices. The customers' base in Northern America is already far over 100 million subscribers and increasing sharply.

Not surprisingly, at the 2012 world radio conference (WRC 2012), African and Middle East countries pushed a motion requiring that, in Region 1, the 700 MHz band be allocated to mobile services, i.e. LTE, like in the USA. European delegations were not aware of this initiative and had to follow the movement.

In Europe, at last the 800 MHz band has been freed for LTE, and the take off of LTE may be expected for the next five years. With around 20 million subscribers, LTE is far behind GSM and UMTS, considering the relative penetration rate. It will probably wait for 2015 when the next WRC 2015 will definitively allow the 700 MHz worldwide to LTE/4G. Already now, LTE is offered in the main European countries, such as the United Kingdom, Germany, France, Italy, Spain Belgium and Switzerland.

In Europe, frequencies for LTE in the 800 MHz band are not optimal: while LTE allows us to engineer LTE with bandwidths from 1.4 MHz to 20 MHz, the allocations are limited to 5 MHz or 10 MHz. Of course, two allocations of 10 MHz, not adjacent, will carry less than one of 20 MHz and the ongoing proposals for the 700 MHz band do not seem to provide large bandwidths.

Let us recall that "LTE Advanced" is supposed to receive 2 × 100 MHz in order to reach 1 Gbps downlink.

## I.2. High speed broadband mobile services: what the customers are waiting for

### I.2.1. *Customers' expectancies*

Demands for wireless data services are showing rapid growth due to evolved networks for high-speed connectivity, wide-scale deployment, flat-rate pricing plans and Internet-friendly devices (smartphones). Consumers rely heavily, and often exclusively, on mobile devices for their communications needs. Therefore, the normal trend is to require, from the mobile system, the same performances as the one offered by fixed networks with ADSL. Very high bit-rate DSL (VDSL), fiber optics or coaxial cable. This comparison raises the level of the bitrate upto 10 Mbps in the first step, and increases upto 30 Mbps. Officially, the target stands at 100 Mbps, the requirement assigned by ITU-T IMT Advanced, but as observed on the fixed networks, very few customers can make a proper use of such a bitrate.

Applications are developed to follow the technical improvement of the systems. They offer a whole range of services, which subsequently increases the request for more bandwidth and more capacity. Basically, they are composed of:

– Internet applications, as for the fixed networks, including mail, downloads and interactive services; this covers laptops, PDAs and fixed broadband services: the most intuitive set of services that can be provided are related to all the fixed wired Digital Subscriber Line (DSL) Internet services that we have today, except that they should be provided wireless and should support mobility;

– multimedia uploads and exchange services. The high uplink data rates of LTE allow for multimedia upload and exchange services such as file sharing, mobile blogging, social networking etc;

– Internet applications specially designed for the mobile user, in particular location based services. The high data rates combined with mobility of LTE spurs a growth in development of newer and better consumer electronic goods leveraging these advantages. Better gaming consoles, vehicular entertainment systems, portable multimedia players, digital cameras with network capabilities and the likes will be introduced, which will add value to the technology;

– television, especially download of movies; and real-time television needing some 4 Mbps or 5 Mbps with H264 or H265 encoding. In this category are premium video on demand/music on demand (VOD/MOD) services. LTE provides effective high data rates and differentiated QoS services. Operators can provide premium multimedia-based services such as VOD and MOD to subscribers who wish to avail such services. The critical point for these services will be superior quality coupled with ease of mobility;

– and of course, telephony, with the possibility of wide band telephony (7 kHz instead of 4 kHz). It will support business applications for vertical markets. LTE allows operators to provide services to vertical business markets through business applications such as video conferencing to enterprise customers, video surveillance, services to homes. The list of services that can be provided through, is only restricted by our imagination. Limitless applications can be supported through a truly mobile broadband infrastructure.

Whichever are the services, wireless operators must also provide a high-quality cellular coverage anywhere customers want to communicate. This requirement is not related to broadband mobile services, it is the principal need for any mobile subscriber and for any service to be provided.

Due to the high costs of backhaul, alternative means to improve cellular coverage in locations, which are difficult to reach, as well as to off-load traffic from the wireless networks. A way to fit to the subscribers' wishes is to install femtocells, taking advantage of the home Internet high-speed link. It is a way to better support residential and small/home office applications. Vodafone UK was the first operator to launch a commercial femtocell service in Europe (July 2009). AT&T (2H 2009) and Verizon (early 2010) also launched commercial femtocell offerings.

From a competitive perspective, femtocells can help mobile operators seize residential minutes from fixed providers, increase market share and respond to emerging Voice over Internet Protocol (VoIP) and Wi-Fi offerings. This of course implies a sharing agreement to be negotiated with the Internet service provider.

From a QoS perspective, femtocells will improve the user experience in the home. This is essential for reducing churn and providing new revenues. Just recall that with the advent of smartphones, mobile communications are heavily using the Internet and high bitrates.

A rapid increase of mobile data usage and the emergence of new applications such as Multimedia Online Gaming (MMOG), mobile TV, web 2.0, streaming contents have motivated the 3GPP to work on the LTE on the way toward 4G mobile.

### I.2.2. *Advantages of LTE for fulfilling these expectancies*

The main goal of LTE is to provide a high data rate, low latency and packet optimized radio access technology supporting flexible bandwidth deployments. At the same time its network architecture has been designed with the goal to support packet-switched traffic with seamless mobility and great QoS.

LTE provides:

– *High throughput:* high data rates can be achieved in both downlink as well as uplink. This causes high throughput.

– *Low latency:* time required to connect to the network is in the range of a few hundred milliseconds and power saving states can now be entered and exited very quickly.

– *FDD and TDD in the same platform:* FDD and Time Division Duplex (TDD), both schemes can be used on same platform.

– *Superior end-user experience:* optimized signaling for connection establishment and other air interface and mobility management procedures have further improved the user experience. Reduced latency (to 10 ms) for better user experience.

– *Seamless Connection:* LTE will also support seamless connection to existing networks such as GSM, CDMA and WCDMA.

– *Plug and play:* the user does not have to manually install drivers for the device. Instead the system automatically recognizes the device, loads new drivers for the hardware if needed and begins to work with the newly connected device.

– *Simple architecture:* because of simple architecture low operating expenditure (OPEX).

### I.2.3. *How the advent of smartphones impacts customers' expectations*

In recent years, the revolutionary event has been the introduction of the iPhone on the mobile market. Earlier, the mobile industry was under the constraints of operators, due to the common practice of operators buying millions of mobiles and including their delivery to the subscriber within the monthly subscription bill, especially in Europe. By these means, they have been able to banish many of the services, which the customer was very keen to obtain. Such applications were relatively easy to include in high-end mobiles, technically speaking.

With the iPhone, Steve Jobs introduced a different paradigm. This paradigm has been the same as the one underlying the phenomenal success of "Minitel" in France. Developers are free to post applications into a common store – such as the "Applestore", managed by Apple. Apple collects the fees from the customers and pays back a certain percentage to the author. In that value chain, the operator is limited to provision of the telecommunication duct and receives little money for the use of its network.

Of course, operators adapted themselves to the new framework. They are now selling iPhones the same way as the other mobile terminals.

Following the path opened by Apple, Google introduced Android, mainly based on Linux software, opened to any manufacturer without fee. As a result, Android is now the dominant standard for smartphones. Microsoftand Blackberry show little success in their smartphones at present. The Android world offers nearly the same applications as the Apple world.

Among thousands of applications, it seems that location services and location based services are the key services. For this purpose, the smartphones include a GPS receiver and the necessary processor of the satellite signals, combined with precise maps of different areas of interest.

However, smartphones include a Wi-Fi access, which is generally put as a priority choice. When Wi-Fi is present, the smartphone will automatically try to connect via the Wi-Fi, instead of the mobile network.

Other successful applications are all kinds of games.

# 1

# LTE Roll-Out

## 1.1. LTE air interface superior features

The LTE E-UTRA is an essential subsystem, so that an optimized packet-based access system can achieve the expected system performance in terms of high data rates and low latency. E-UTRA is also expected to support mobility up to 350 km/h at least at certain frequency values, conserve mobile station's power consumption through microsleep and provide seamless integration of unicast and enhanced broadcast transmission. The key techniques for the long-term evolution (LTE) air interface are summarized below.

### 1.1.1. *Orthogonal frequency division multiplexing access (OFDMA) for the downlink*

OFDMA allows data to be transmitted in parallel in a set of narrowband, orthogonal and tightly packed subcarriers, providing an efficient use of the available bandwidth. The use of cyclic prefix in OFDMA makes it robust to time-dispersion (multipath) without the need for complex equalizers at the receiver end, which reduces complexity, cost and power consumption.

### 1.1.2. *Single-carrier frequency division multiple access for uplink*

OFDMA produces large output variations and requires highly linear power amplifiers that inherently have a low power efficiency.

Since power consumption is extremely important for the UE, plain OFDMA is not used for the uplink but a DFT-precoded OFDM, also known as single-carrier OFDMA (SC-FDMA). SC-FDMA comes as a power efficient alternative of OFDMA that retains most of the advantages of OFDMA.

### 1.1.3. *Multiple-input multiple-output (MIMO) transmission*

Two major limitations in communications channels can be multipath interference, and the data throughput limitations as a result of Shannon's law. MIMO provides a way of utilizing the multiple signal paths that exist between a transmitter and receiver to significantly improve the data throughput available on a given channel with its defined bandwidth. By using multiple antennas at the transmitter and receiver along with some complex digital signal processing, MIMO technology enables the system to set up multiple data streams on the same channel, thereby increasing the data capacity of a channel.

MIMO is being used increasingly in many high data rate technologies, including Wi-Fi and other wireless and cellular technologies, to provide improved levels of efficiency. Essentially, MIMO employs multiple antennas on the receiver and transmitter to utilize the multipath effects that always exist to transmit additional data, rather than causing interference.

#### 1.1.3.1. *MIMO in LTE*

MIMO techniques enhance system performance, service capabilities or both. At its highest level, LTE multiantenna transmission can be divided into transmit diversity and spatial multiplexing. The former can be seen as a technique for averaging the signals received from the two antennas, thereby avoiding the deep fading dips that occur per antenna. The latter employs multiple antennas at the transmitter and receiver side to provide simultaneous transmission of multiple parallel data streams over a single radio link, therefore, increasing significantly the peak data rates over the radio link. Additionally, LTE supports spatial division multiple access (SDMA) and beamforming.

Downlink MIMO configuration reaches up to an $8 \times 8$ antenna system, supports MU-MIMO and offers enhancements to CSI feedback.

Uplink MIMO introduces UL transmit diversity. Antenna configuration: up to $4 \times 4$ SU-MIMO.

MIMO is used within LTE to provide better signal performance and/or higher data rates by the use of the radio path reflections that exist.

MIMO is another form of the major LTE technological innovations used to improve the performance of the system. This technology provides LTE with the ability to further improve its data throughput and spectral efficiency above that obtained by the use of OFDM.

Although MIMO adds complexity to the system in terms of processing and the number of antennas required, it enables far higher data rates to be achieved along with much improved spectral efficiency. As a result, MIMO has been included as an integral part of LTE.

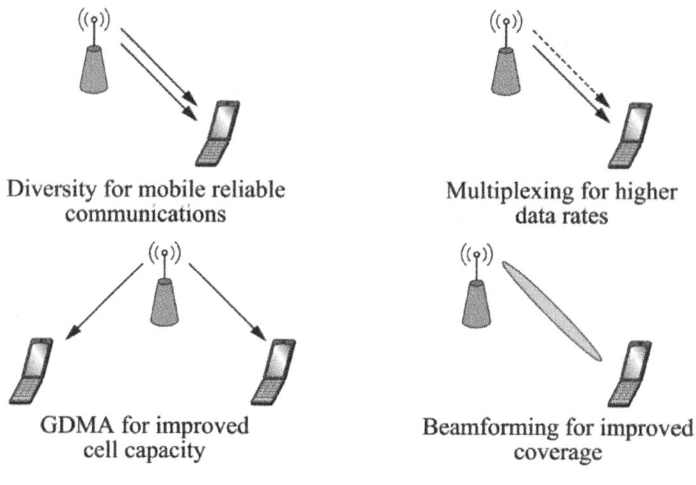


**Figure 1.1.** *Use of MIMO techniques in LTE*

1.1.3.2. *LTE MIMO basics*

The basic concept of MIMO utilizes the multipath signal propagation that is present in all terrestrial communications.

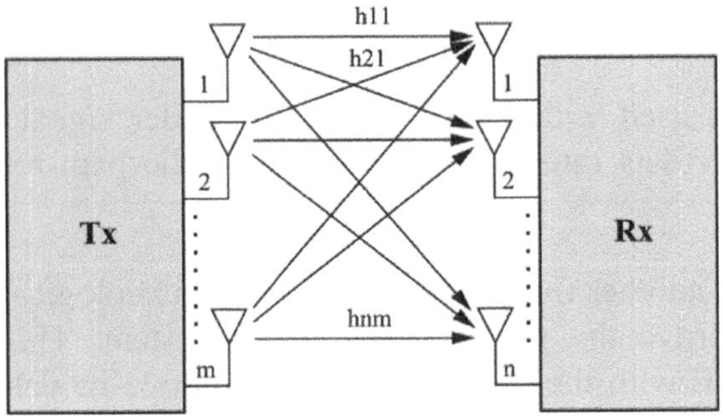


**Figure 1.2.** *General outline of MIMO system*

The transmitter and receiver have more than one antenna and using the processing power available at either end of the link, they are able to utilize the different paths that exist between the two entities to provide improvements in the data rate of signal to noise.

The use of MIMO technology has been introduced successively over the different releases of the LTE standards.

MIMO has been a cornerstone of the LTE standard, but initially, in releases 8 and 9 multiple transmit antennas on the UE were not supported because, in the interest of power reduction, only a single radiofrequency (RF) power amplifier was assumed to be available. It was in release 10 that a number of new schemes were introduced. Closed loop spatial multiplexing for SU-MIMO as well as multiple antennas on the UE.

1.1.3.3. *LTE MIMO modes*

There are several ways in which MIMO is implemented in LTE. These vary according to the equipment used, the channel function and the equipment involved in the link.

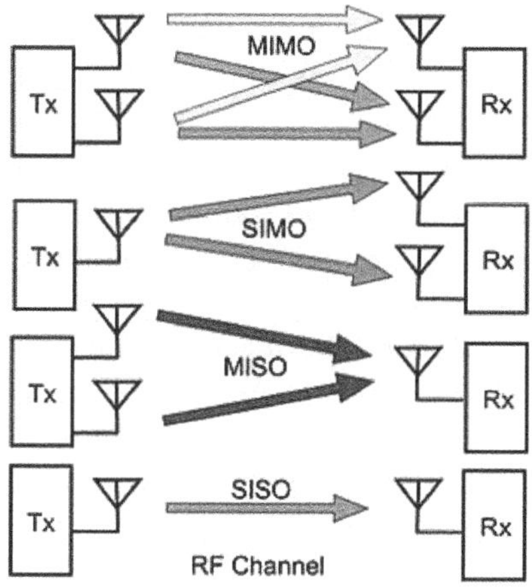


**Figure 1.3.** *Tx Rx*

*Single antenna:* this is the form of wireless transmission used on most basic wireless links. A single data stream is transmitted on one antenna and received by one or more antennas. It may also be referred to as single in single out (SISO) or single in multiple out (SIMO) dependent upon the antennas used. SIMO is also called receive diversity.

*Transmit diversity:* this form of LTE MIMO scheme utilizes the transmission of the same information stream from multiple antennas. LTE supports two or four for this technique. The information is coded differently using space frequency block codes. This mode provides an improvement in signal quality at reception and does not improve the data rate. Accordingly, this form of LTE MIMO is used on the common channels as well as the control and broadcast channels.

*Open loop spatial multiplexing:* this form of MIMO used within the LTE system involves sending two information streams that can be transmitted over two or more antennas. However, there is no feedback from the UE although a transmit rank indicator (TRI) transmitted from the UE can be used by the base station to determine the number of spatial layers.

*Close loop spatial multiplexing:* this form of LTE MIMO is similar to the open loop version, but as the name indicates it has feedback incorporated to close the loop. A precoding matrix indicator (PMI) is fed back from the UE to the base station. This enables the transmitter to precode the data to optimize the transmission and enables the receiver to more easily separate the different data streams.

*Closed loop with precoding:* this is another form of LTE MIMO, but where a single code word is transmitted over a single spatial layer. This can be sued as a fall-back mode for closed loop spatial multiplexing, and it may also be associated with beamforming as well.

*Multiuser MIMO, MU-MIMO:* this form of LTE MIMO enables the system to target different spatial streams to different users. In order to meet the peak spectrum efficiency, antenna configurations of $8 \times 8$ for downlink transmission and $4 \times 4$ for uplink transmission are being investigated. Furthermore, LTE-A MIMO technologies are also designed with the aim of improving cell average throughput as well as cell edge performance. A uniform and adaptive MIMO platform is considered in order to accommodate the demand of high data rates and wider coverage by switching from one mode to another.

#### 1.1.3.4. *Beamforming*

This is the most complex of the MIMO modes, and it is likely to use linear arrays that will enable the antenna to focus on a particular area. This will reduce interference and increase capacity as the particular UE will have a beam formed in its particular direction. In this, a single code word is transmitted over a single spatial layer. A dedicated reference signal is used for an additional port. The terminal estimates the channel quality from the common reference signals on the antennas. Beyond the increased number of antennas, single-site MIMO evolutions for LTE-A also include an adaptive strategy regarding the beamforming approach that needs to be investigated. Depending on the mobility, antenna configuration and cell size, a fixed-beam (e.g. grid-of-beams, GoB) or a user-specific-beam technique could be selected. In addition, preference for coverage requirements versus peak rates led to a new transmission mode where beamforming is combined with spatial multiplexing within different beams.

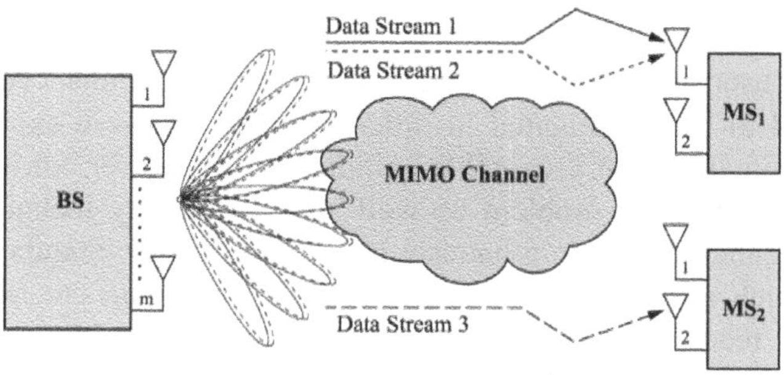


**Figure 1.4.** *Beamforming and spatial multiplexing combination*

1.1.3.5. *Multisite MIMO*

Multisite MIMO is a novel approach in LTE-A that seeks to improve the cell edge performance by means of spatial multiplexing from different base stations that share the same spectral resources. In the downlink, two different versions of multisite MIMO are defined, namely network MIMO, where base stations share coherent short-term channel information, and collaborative MIMO, where non-coherent long-term channel information is shared.

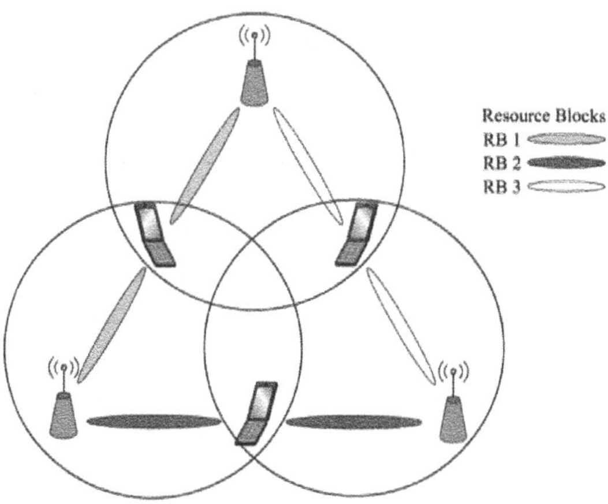


**Figure 1.5.** *Multisite MIMO scenario*

In the uplink, however, an application of network MIMO coherently coordinates a reasonable number of base stations in reception. This facilitates interference reduction among multiple bases that must compute beamforming weights to maximize SINR values for each user. Many are the challenges regarding enhanced MIMO that need to be investigated in order to fulfill all these expectations while maintaining an acceptable complexity and power consumption. The physical space problem as well as the diversity techniques for eight antennas placed in a handset have to be solved. Standardization issues, such as design of reference signals, signaling, etc. and more complex feedback schemes, have to be studied. Additionally, the question of when a terminal is eligible for coordinated transmission needs to be addressed, since the tradeoff between cell average throughput and cell edge performance has to be optimized.

#### 1.1.3.6. *Coordinated multiple point (CoMP) transmission and reception*

LTE-A defines in general terms CoMP as the "coordination in the downlink/uplink from/to multiple geographically separated transmission/reception points". Antennas of multiple cell sites are used in such a way that they can contribute to improve the quality of the received signal at the UE/eNB and drastically reduce the intercell interference. This will demand very fast inter-eNB connections and some additional control strategies that might be centralized or not. There are mainly two types of CoMP in the downlink that differ in the degree of coordination. They are presented below.

Regarding the uplink, CoMP techniques are less advanced due to the impossibility of ensuring the connectivity and data sharing among terminals. Data are received at multiple base stations, and scheduling is coordinated in order to reduce interference. The receiving base stations must incorporate some signal processing techniques to process the different streams. We will study the challenges and propose possible solutions for CoMP transmission/reception for scheduling and joint transmission/reception in scenarios with/without MIMO technology at the UE/eNodeB. More importantly, investigation is necessary to check if the increased complexity of these techniques is compensated with the achievable improved performance.

### 1.1.3.7. *Coordinated scheduling/beamforming*

In this case, data are only transmitted from a single eNB, but base stations are connected with each other in order to exchange scheduling and beamforming information so that a dynamic multisite scheduling can be performed. The requirements concerning synchronization among base stations and backhaul capacity are obviously lower.

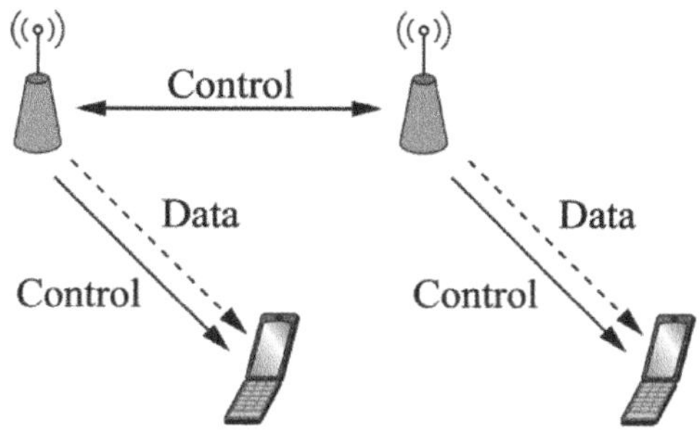


**Figure 1.6.** *Control and data*

### 1.1.3.8. *Joint processing*

Data are transmitted from different base stations at the same time, therefore, requiring a tight synchronization and a very high-speed link among base stations. Two techniques are possible: fast cell selection, where only one base station is transmitting at a time, and joint transmission where data are transmitted from different points at one time and they are coherently combined at the terminal.

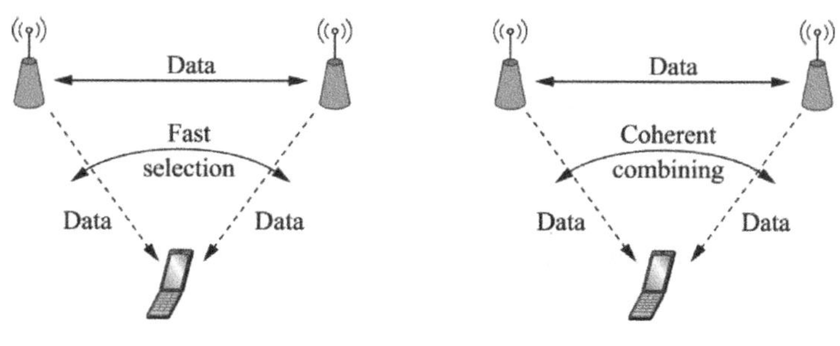


**Figure 1.7.** *Joint processing techniques*

### 1.1.4. *Support for component carrier*

Today, the maximum bandwidth of LTE is 20 MHz. LTE-A will support up to 100 MHz bandwidth by aggregating two or more LTE "Component Carriers" (CCs) of up to 20 MHz. These CCs can be continuous or discontinuous in a single spectrum band, or from different spectrum bands.

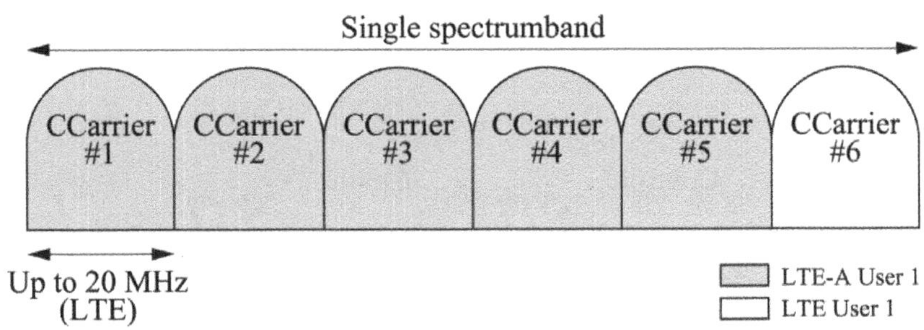


**Figure 1.8.** *Bandwidth aggregation in contiguous bandwidth, single spectrum band*

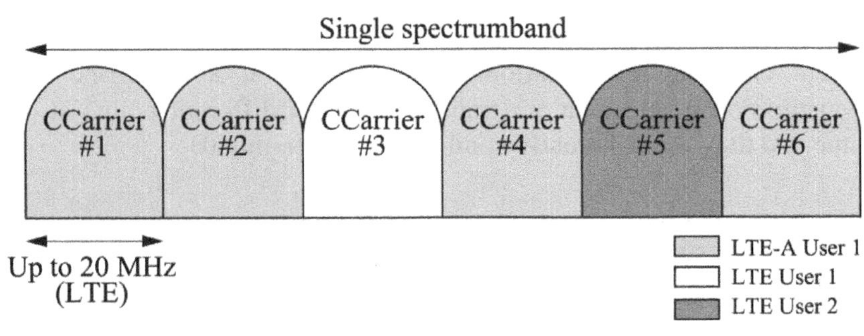


**Figure 1.9.** *Bandwidth aggregation in non-contiguous bandwidth, single spectrum band*

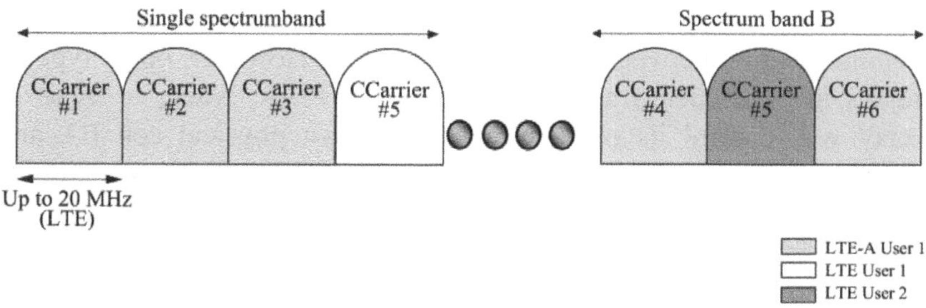


**Figure 1.10.** *Bandwidth aggregation in non-contiguous bandwidth, multiple spectrum bands*

Several challenges exist to achieve high utilization, with low cost/complexity, of these scenarios. We will study these challenges and propose possible solutions, starting from the lowest layers of the protocol stack to the upper ones, taking into account different elements of integration such as: transceiver design, resource assignment based on user and system requirements, hand-over procedures, error control and transport protocol optimizations, among others.

### 1.1.5. *Relaying*

In order to improve the coverage of high data rates, group mobility, temporary network deployment, cell-edge throughput, and to provide coverage in new areas, LTE-A includes support for relays. The basic architecture analyzed for LTE-A consists of a single relay node (RN) that is connected to a *donor cell* of a *donor eNodeB*.

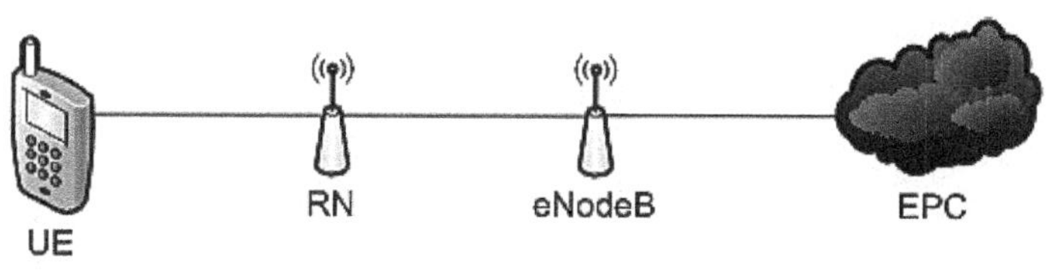


**Figure 1.11.** *Basic relay scheme for LTE-A*

LTE-A is considered for the use of inband and outband communication between the RN and eNodeB. However, only "type 1" relays are considered as a minimum for LTE-A, where a "type 1" relay will control its own cell, with its own physical cell ID, and manage its own scheduling, error control and UE feedback. For an LTE device, an RN will appear as an eNodeB. Beyond this type of relay, there are several other possible relay "types" that could be used. For example:

– Transparent versus non-transparent: a transparent relay will appear as another multipath to the UE, while a non-transparent will appear as a new entity (an eNodeB or an RN) to the UE.

– Half-duplex versus full-duplex: a half-duplex relay can only transmit or receive at a time instant, while a full-duplex relay can transmit and receive simultaneously.

– Single-antenna versus multiantenna relay: an RN could manage one or multiple antennas to transmit/receive to/from the UE/eNodeB.

– L1 versus L2 versus L3 relay: an L1 relay will simply forward all received signals, an L2 relay will include a certain degree of processing before retransmitting the received signals (decode, error correction, etc.) and an L3 relay will be an eNodeB that is acting as a relay station.

– Coordination versus non-coordination relays: an RN can work individually, or cooperate/coordinate, similarly to the eNodeB to achieve similar benefits of CoMP.

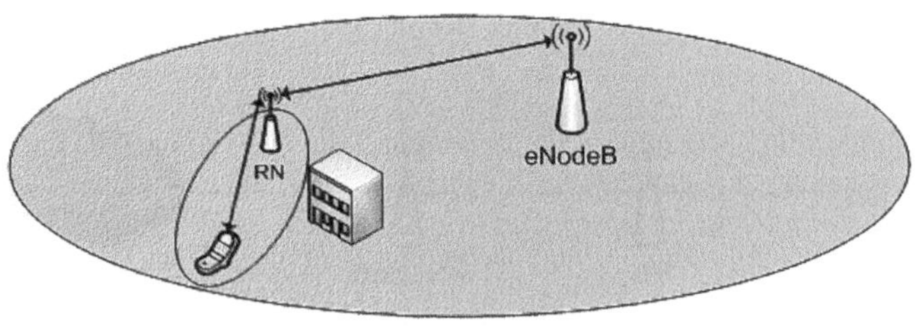


**Figure 1.12.** *RN improving coverage*

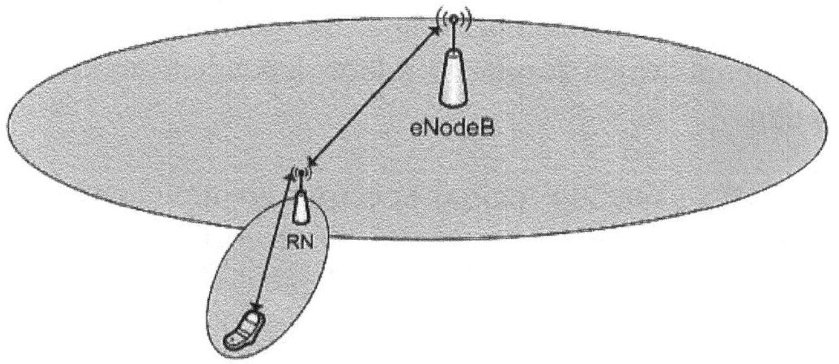


**Figure 1.13.** *RN extending coverage*

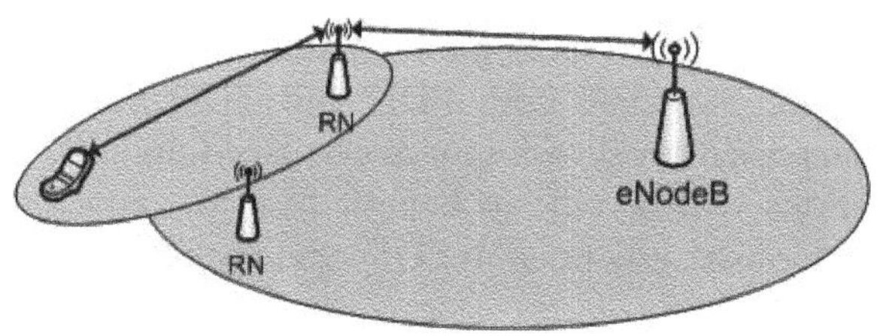


*RN with no cooperation/coordination*

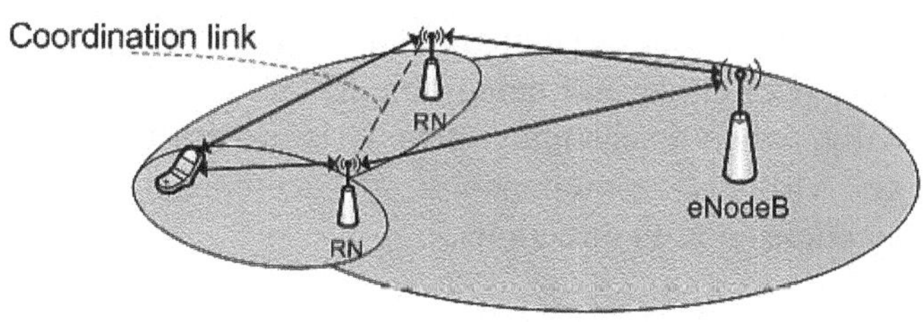


**Figure 1.14.** *RN with cooperation/coordination*

## 1.2. LTE FDD, TDD and TD-LTE duplex schemes

LTE has been defined to accommodate both paired spectrum for frequency division duplex, FDD, and unpaired spectrum for time division duplex, TDD, operation. It is anticipated that both LTE TDD and LTE FDD will be widely deployed as each form of the LTE standard has its own advantages and disadvantages, and decisions can be made about which format to adopt dependent upon the particular application.

LTE FDD using the paired spectrum is anticipated to form the migration path for the current 3G services being used around the globe, most of which use FDD-paired spectrum. However, there has been an additional emphasis on including TDD LTE using unpaired spectrum. TDD LTE, which is also known as TD-LTE, is seen as providing the evolution or upgrade path for TD-SCDMA.

In view of the increased level of importance being placed upon LTE TDD or TD-LTE, it is planned that user equipment will be designed to accommodate both FDD and TDD modes. With TDD having an increased level of importance placed upon it, it means that TDD operations will be able to benefit from the economies of scale that were previously only open to FDD operations.

### 1.2.1. *Duplex schemes*

It is essential that any cellular communications system be able to transmit in both directions simultaneously. This enables conversations to be made, with either end being able to talk and listen as required. Additionally, when exchanging data it is necessary to be able to undertake virtually simultaneous or completely simultaneous communications in both directions.

It is necessary to be able to specify the different directions of transmission so that it is possible to easily identify in which direction the transmission is being made. There are a variety of differences between the two links ranging from the amount of data carried to the transmission format, and the channels implemented. The two links are defined as:

– *Uplink:* the transmission from the UE, or user equipment, to the eNodeB or base station.

– *Downlink*: the transmission from the eNodeB or base station to the UE.

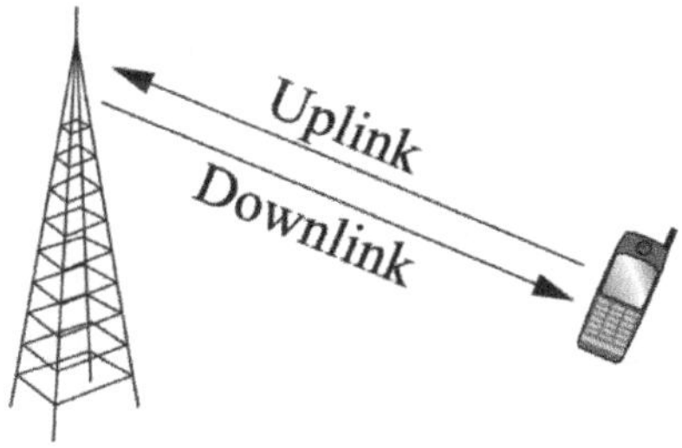


**Figure 1.15.** *Uplink and downlink transmission directions*

In order to be able to transmit in both directions, the UE or base station must have a duplex scheme. There are two forms of duplex that are commonly used, namely FDD and TDD.

In order for radio communications systems to be able to communicate in both directions, it is necessary to have what is termed a duplex scheme. A duplex scheme provides a way of organizing the transmitter and receiver so that they can transmit and receive. There are several methods that can be adopted. For applications including wireless and cellular telecommunications, where it is required that the transmitter and receiver are able to operate simultaneously, two schemes are in use. One is known as FDD and uses two channels, one for transmitting and the other for receiving. Another scheme known as TDD uses one frequency but allocates different time slots for transmission and reception.

Both FDD and TDD have their own advantages and disadvantages. Accordingly, they may be used for different applications, or where the bias of the communications is different.

*Advantages/disadvantages of LTE TDD and LTE FDD for cellular communications*

There are a number of advantages and disadvantages of TDD and FDD that are of particular interest to mobile or cellular telecommunications operators. These are naturally reflected in LTE.

| COMPARISON OF TDD LTE AND FDD LTE DUPLEX FORMATS | | |
|---|---|---|
| PARAMETER | LTE-TDD | LTE-FDD |
| Paired spectrum | Does not require paired spectrum as both transmit and receive occur on the same channel | Requires paired spectrum with sufficient frequency separation to allow simultaneous transmission and reception |
| Hardware cost | Lower cost as no diplexer is needed to isolate the transmitter and receiver. As cost of the UEs is of major importance because of the vast numbers that are produced, this is a key aspect | Diplexer is needed and cost is higher |
| Channel reciprocity | Channel propagation is the same in both directions that enables transmit and receive to use on set of parameters | Channel characteristics different in both directions as a result of the use of different frequencies |
| UL/DL asymmetry | It is possible to dynamically change the UL and DL capacity ratio to match demand | UL/DL capacity determined by frequency allocation set out by the regulatory authorities. It is, therefore, not possible to make dynamic changes to match capacity Regulatory changes would normally be required, and capacity is normally allocated so that it is the same in either direction |
| Guard period/guard band | Guard period required to ensure that uplink and downlink transmissions do not clash. Large guard period will limit capacity. Larger guard period normally required if distances are increased to accommodate larger propagation times | Guard band required to provide sufficient isolation between uplink and downlink. Large guard band does not impact capacity |

| | | |
|---|---|---|
| Discontinuous transmission | Discontinuous transmission is required to allow both uplink and downlink transmissions. This can degrade the performance of the RF power amplifier in the transmitter | Continuous transmission is required |
| Cross slot interference | Base stations need to be synchronized with respect to uplink and downlink transmission times. If neighboring base stations use different uplink and downlink assignments and share the same channel, then interference may occur between cells | Not applicable |

**Table 1.1.** *TDD LTE AND FDD LTE duplex format parameter*

### 1.2.2. *LTE TDD/TD-LTE and TD-SCDMA*

Apart from the technical reasons and advantages of using LTE TDD/TD-LTE, there are market drivers as well. With TD-SCDMA now well-established in China, there needs to be a 3.9G and later a 4G successor to the technology. With unpaired spectrum allocated for TD-SCDMA as well as UMTS TDD, it is natural to see many operators wanting an upgrade path for their technologies to benefit from the vastly increased speeds and improved facilities of LTE. Accordingly, there is considerable interest in the development of LTE TDD, which is also known as TD-LTE in China.

With the considerable interest from the supporters of TD-SCDMA, a number of features to make the mode of operation of TD-LTE more of an upgrade path for TD-SCDMA have been incorporated. One example of this is the subframe structure that has been adopted within LTE TDD/TD-LTE.

While both LTE TDD (TD-LTE) and LTE FDD will be widely used, it is anticipated that LTE FDD will be the more widespread, although LTE TDD has a number of significant advantages, especially in terms of higher spectrum efficiency that can be used by many operators. It is also anticipated that phones will be able to operate using either the LTE FDD or LTE-TDD (TD-LTE) modes. In this way, the LTE UEs will be dual standard phones, and able to operate in countries regardless of the flavor of LTE that is used – the main problem will then be the frequency bands that the phone can cover.

### 1.2.3. *FDD LTE frequency band allocations*

There is a large number of allocations or radio spectrum that have been reserved for FDD, frequency division duplex, and LTE use.

The FDD LTE frequency bands are paired to allow simultaneous transmission on two frequencies. The bands also have a sufficient separation to enable the transmitted signals not to unduly impair the receiver performance. If the signals are too close then the receiver may be "blocked" and the sensitivity impaired. The separation must be sufficient to enable the roll-off of the antenna filtering to give sufficient attenuation of the transmitted signal within the receive band.

| FDD LTE BANDS & FREQUENCIES | | | | | |
|---|---|---|---|---|---|
| LTE BAND NUMBER | UPLINK (MHZ) | DOWNLINK (MHZ) | WIDTH OF BAND (MHZ) | DUPLEX SPACING (MHZ) | BAND GAP (MHZ) |
| 1 | 1,920–1,980 | 2,110–2,170 | 60 | 190 | 130 |
| 2 | 1,850–1,910 | 1,930–1,990 | 60 | 80 | 20 |
| 3 | 1,710–1,785 | 1,805–1,880 | 75 | 95 | 20 |
| 4 | 1,710–1,755 | 2,110–2,155 | 45 | 400 | 355 |
| 5 | 824–849 | 869–894 | 25 | 45 | 20 |
| 6 | 830–840 | 875–885 | 10 | 35 | 25 |
| 7 | 2,500–2,570 | 2,620–2,690 | 70 | 120 | 50 |
| 8 | 880–915 | 925–960 | 35 | 45 | 10 |
| 9 | 1,749.9–1,784.9 | 1,844.9–1,879.9 | 35 | 95 | 60 |

| 10 | 1,710–1,770 | 2,110–2,170 | 60 | 400 | 340 |
|---|---|---|---|---|---|
| 11 | 1,427.9–1,452.9 | 1,475.9–1,500.9 | 20 | 48 | 28 |
| 12 | 698–716 | 728–746 | 18 | 30 | 12 |
| 13 | 777–787 | 746–756 | 10 | –31 | 41 |
| 14 | 788–798 | 758–768 | 10 | –30 | 40 |
| 15 | 1,900–1,920 | 2,600–2,620 | 20 | 700 | 680 |
| 16 | 2,010–2,025 | 2,585–2,600 | 15 | 575 | 560 |
| 17 | 704–716 | 734–746 | 12 | 30 | 18 |
| 18 | 815–830 | 860–875 | 15 | 45 | 30 |
| 19 | 830–845 | 875–890 | 15 | 45 | 30 |
| 20 | 832–862 | 791–821 | 30 | –41 | 71 |
| 21 | 1,447.9–1,462.9 | 1,495.5–1,510.9 | 15 | 48 | 33 |
| 22 | 3,410–3,500 | 3,510–3,600 | 90 | 100 | 10 |
| 23 | 2,000–2,020 | 2,180–2,200 | 20 | 180 | 160 |
| 24 | 1,625.5–1,660.5 | 1,525–1,559 | 34 | –101.5 | 135.5 |
| 25 | 1,850–1,915 | 1,930–1,995 | 65 | 80 | 15 |
| 26 | 814–849 | 859–894 | 30/40 | | 10 |
| 27 | 807–824 | 852–869 | 17 | 45 | 28 |
| 28 | 703–748 | 758–803 | 45 | 55 | 10 |
| 29 | n/a | 717–728 | 11 | | |
| 30 | 2,305–2,315 | 2,350–2,360 | 10 | 45 | 35 |
| 31 | 452.5–457.5 | 462.5–467.5 | 5 | 10 | 5 |

**Table 1.2.** *Bands and frequencies*

### 1.2.4. *Allocated frequency bands in Europe, multiband operation*

In Europe, for the 800 MHz band, the chosen LTE is FDD with the downlink on 791–821 MHz and 832–862 MHz in the uplink. These frequencies were auctioned at a very high price. In France, only three operators out of four bought such frequencies.

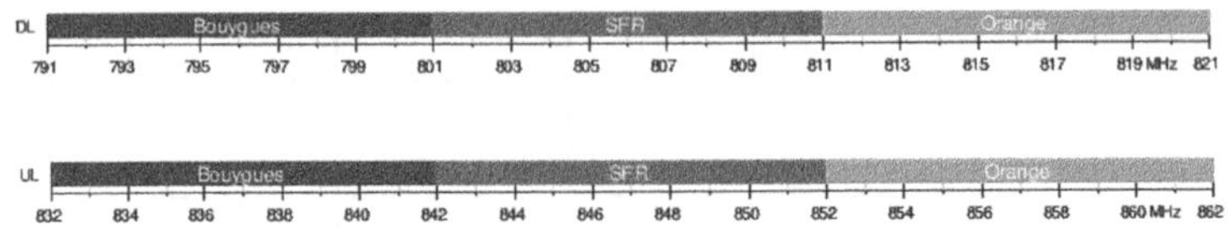


**Figure 1.16.** *Frequencies in France*

In UK, Germany and Italy, this new auction round has been an opportunity to reconsider the whole spectrum that is allocated to mobile communications, especially the GSM frequencies (880–915 and 925–960) and the ex-DCS frequencies (1,710.2–1,784.8 and 1,805.2–1,879.8). Since the US is also deploying LTE in the latter, so-called 1,800 MHz band, there is an opportunity to adopt mobile terminals coming from the massive production serving North America. Hence, the choice by Bouygues Telecom is to implement LTE in this frequency band modified after 26 May 2016 as in Figure 1.17.

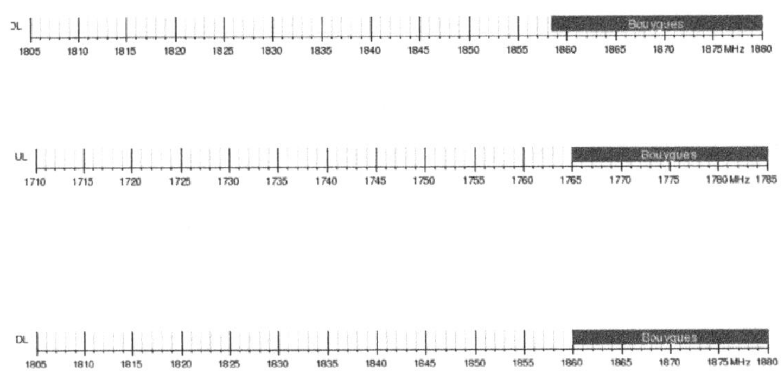


**Figure 1.17.** *Frequencies in France*

All European countries set a combined auction with 800 MHz (the "golden frequencies") that provides an excellent propagation and savings in the coverage building and 2,600 MHz, quasi-Wi-Fi frequencies, ensuring a difficult engineering for the coverage of wide areas. From the wideband code division multiple access (WCDMA) experience in France, to ensure a correct commercial coverage needs more than three times the number of cell sites at 2,100 MHz compared with the same service at 900 MHz.

The 2,600 MHz band has been allocated in France as depicted in Figure 1.18.

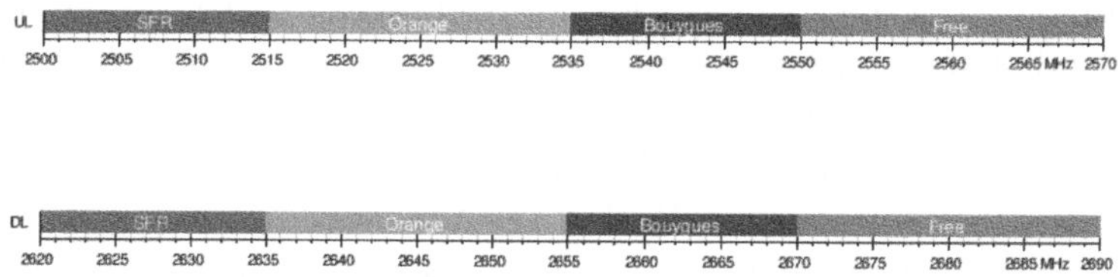


**Figure 1.18.** *Frequencies in France*

For 2,600 MHz, the 4 h French operator has bought some frequencies, which were considerably cheaper. Its idea is probably to equip urban areas quickly and wait for 700 MHz to cover the rural areas.

### 1.2.5. *TDD LTE frequency band allocations*

With the interest in TDD LTE, there are several unpaired frequency allocations that are being prepared for LTR TDD use. The TDD LTE bands are unpaired because the uplink and downlink share the same frequency, being time multiplexed.

| TDD LTE BANDS & FREQUENCIES | | |
|---|---|---|
| LTE BANDNUMBER | ALLOCATION (MHZ) | WIDTH OF BAND (MHZ) |
| 33 | 1,900–1,920 | 20 |
| 34 | 2,010–2,025 | 15 |
| 35 | 1,850–1,910 | 60 |
| 36 | 1,930–1,990 | 60 |
| 37 | 1,910–1,930 | 20 |
| 38 | 2,570–2,620 | 50 |
| 39 | 1,880–1,920 | 40 |
| 40 | 2,300–2,400 | 100 |
| 41 | 2,496–2,690 | 194 |
| 42 | 3,400–3,600 | 200 |
| 43 | 3,600–3,800 | 200 |
| 44 | 703–803 | 100 |

**Table 1.3.** *TDD LTE bands and frequencies*

There are regular additions to the LTE frequency bands/LTE spectrum allocations as a result of negotiations at the ITU regulatory meetings. These LTE allocations are resulting in part from the digital dividend, and also from the pressure caused by the ever growing need for mobile communications. Many of the new LTE spectrum allocations are relatively small, often 10–20 MHz in bandwidth, and this is a cause for concern. With LTE-Advanced needing bandwidths of 100 MHz, channel aggregation over a wide set of frequencies may be needed, and this has been recognized as a significant technological problem.

## 1.3. LTE UE category and class definitions

LTE utilizes UE categories or classes to define the performance specifications and enable base stations to be able to communicate effectively with them knowing their performance levels.

In the same way that category information is used for virtually all cellular systems from GPRS onward, so the LTE UE category information is of great importance. While users may not be particularly aware of the category of their UE, it will match the performance and allow the eNB to communicate effectively with all the UEs that are connected to it.

In the same way that a variety of other systems adopted different categories for the handsets or UEs, so too are there 3G LTE UE categories. These LTE categories define the standards to which a particular handset, dongle or other piece of equipment will operate.

### 1.3.1. *LTE UE category rationale*

The LTE UE categories or UE classes are needed to ensure that the base station, or eNodeB, eNB can communicate correctly with the user equipment. By relaying the LTE UE category information to the base station, it is able to determine the performance of the UE and communicate with it accordingly.

As the LTE category defines the overall performance and capabilities of the UE, it is possible for the eNB to communicate using capabilities that it knows the UE possesses. Accordingly, the eNB will not communicate beyond the performance of the UE.

**Figure 1.19.** *Smartphone*

### 1.3.2. *LTE UE category definitions*

There are five different LTE UE categories that are defined. As can be seen in Table 1.4, the different LTE UE categories have a wide range in the supported parameters and performance. LTE category 1, for example, does not support MIMO, but LTE UE category 5 supports 4 × 4 MIMO.

It is also worth noting that UE class 1 does not offer the performance offered by that of the highest performance HSPA category. Additionally, all LTE UE categories are capable of receiving transmissions from up to four antenna ports.

A summary of the different LTE UE category parameters is given in Tables 1.4 and 1.5.

| HEADLINE DATA RATES FOR LTE UE CATEGORIES | | | | | |
|---|---|---|---|---|---|
| | CATEGORY | | | | |
| LINK | 1 | 2 | 3 | 4 | 5 |
| Downlink | 10 | 50 | 100 | 150 | 300 |
| Uplink | 5 | 25 | 50 | 50 | 75 |

**Table 1.4.** *Headline data rates for LTE UE categories*

| UL AND DL PARAMETERS FOR LTE UE CATEGORIES | | | | | |
|---|---|---|---|---|---|
| | CATEGORY | | | | |
| PARAMETER | 1 | 2 | 3 | 4 | 5 |
| Maximum number of DL-SCH transport block bits received in a TTI | 10 296 | 51 024 | 102 048 | 150 752 | 302 752 |
| Maximum number of bits of a DL-SCH block received in a TTI | 10 296 | 51 024 | 75 376 | 75 376 | 151 376 |
| Total number of soft channel bits | 250 368 | 1 237 248 | 1 237 248 | 1 827 072 | 3 667 200 |
| Maximum number of supported layers for spatial multiplexing in DL | 1 | 2 | 2 | 2 | 4 |
| Maximum number of bits of an UL-SCH transport block received in a TTI | 5 160 | 25 456 | 51 024 | 51 024 | 75 376 |
| Support for 64-QAM in UL | No | No | No | No | Yes |

**Table 1.5.** *UL and DL parameters for LTE UE categories*

While the headline rates for different LTE UE categories or UE classes show the maximum data rates achievable, it is worth looking in further detail at the underlying performance characteristics.

From this, it can be seen that the peak downlink data rate for a category 5 UE using 4 × 4 MIMO is approximately 300 Mbps, and 150 Mbps for a category 4 UE using 2 × 2 MIMO. Also in the uplink, LTE UE category 5 provides a peak data rate of 75 Mbps using 64-QAM.

*DL-SCH = Downlink shared channel*

*UL-SCH = Uplink shared channel*

*TTI = Transmission Time Interval*

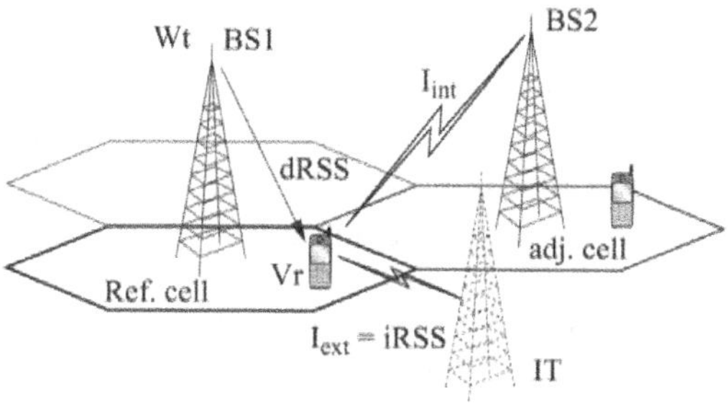


**Figure 1.20.** *Interference mechanism in the OFDMA*

## 1.4. Interferences in OFDMA

The interference created by the system itself is called self-interference $I_{inter}$. The interference coming from another system is called $I_{ext}$.

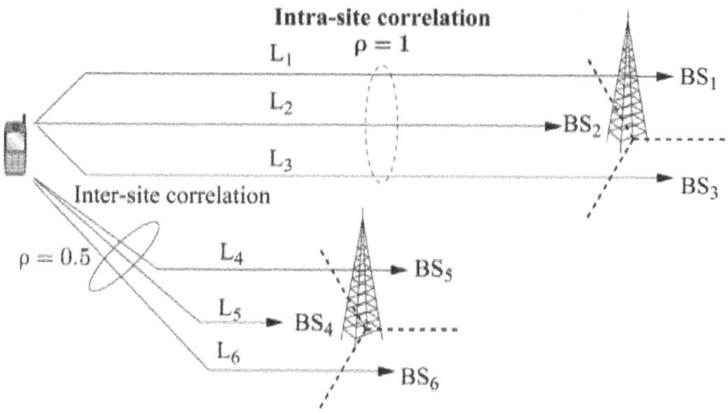


**Figure 1.21.** *Intra-/inter-site correlation*

The panel is similar for the OFDMA DL and UL. The concept of a simple correlation model for shadow fading has been widely adopted in LTE coexistence studies mostly employed in uplink case. The propagation attenuation is modeled as the product of the path loss and shadow fading. The shadow fading is well-approximated by a log-normal distribution.

Let z denote shadow fading in dB with zero mean and variance $\sigma^2$. Then, the shadow fading of path from one UE to the *i-th* BS is expressed as $z_i=a*x + b*y_i$, where $a^2+b^2=1$ and $x$ and $y_i$ are independent Gaussian distributed variables, both with zero mean and variance $\sigma^2$ . $y_i$ and $y_j$ are also independent.

Thus, the correlation coefficient of the shadow fading from one UE to two different BSs, i.e. the *i-th* and *j-th* BS, is:

$$\frac{E(z_i z_j)}{E(z_i^2)} = a^2$$

In most LTE studies,

$$a = b = \frac{1}{\sqrt{2}}$$

is assumed [TR36.942]. For cellular systems with three-sector antennas, the shadowing correlation between sites (equivalent to BS in Omni antenna system) is of 0.5 and correlation between sectors of the same site is consequently of 1.

*Propagation model*

The interferences are related to the propagation. Many propagation model scans have been designed, either with logarithmic formulas (Hata type models) or with ray tracing (mostly for dense urban areas).

*Positioning*

To calculate interferences, the positioning of radiating elements must be very precise. The use of GPS positioning is recommended.

*Transmitter to victim link receiver path*

When OFDMA is an interferer, it is necessary to set the characteristic of the path between the interfering transmitter (UE for UL or BS for DL) and the victim system.

– Calculation of the UE frequencies in UL

The frequency of the UE in UL is calculated as follows:

$$F_{UE} = F_{system} - (BW_{system} / 2) + ((((N_{RB\_UE} * BW_{RB}) + (diff/ N_{UE})) / 2) * ((Index_{link} * 2) + 1))$$

with: $diff = BW_{system} - (N_{RB\_BS} * BW_{RB})$

Diff takes into account for any difference between the $BW_{system}$ and the effective bandwidth (NRB_BS * BWRB):

– $F_{UE}$: center frequency of the UE;

– $F_{system}$: frequency of the system (i.e. the network) (input to SEAMCAT);

– $BW_{system}$: bandwidth of the system (input to SEAMCAT);

– $N_{RB\_UE}$: number of resource blocks (RBs) per mobile (input to SEAMCAT);

– $N_{RB\_BS}$: number of RB for the BS (input to SEAMCAT);

– $BW_{RB}$: bandwidth of the RB (input to SEAMCAT);

– $N_{UE}$: number of UEs in the system (calculated as NRB_BS/NRB_UE);

– $Index_{link}$: index of the specific link UE to serving BS (input to SEAMCAT). Index = [0, NUE-1].

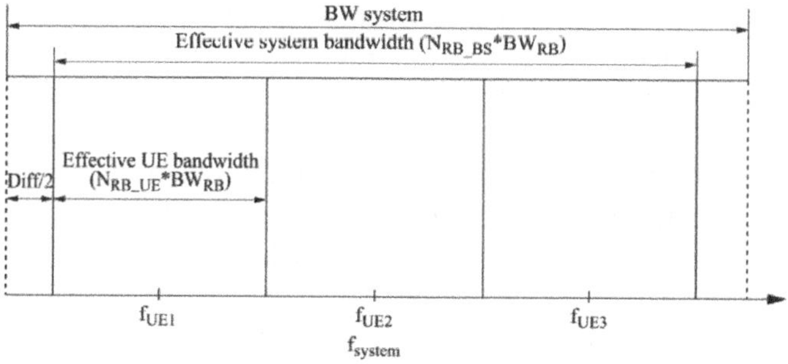


**Figure 1.22.** *Effective system bandwidth*

*OFDMA UL power control*

In OFDMA UL, the power control is applied to the active users (i.e. the mobile users with specific RBs) so that the UE Tx power is adjusted with respect to effective path loss (i.e. based on the MCL) to the BS to which it is connected to. In 3GPP [TR36.942], the UL power control is defined so that the UE transmit power is set such as:

$$P_t = P_{\max} \times \min\left\{1, \max\left[R_{\min,}\left(\frac{PL}{PL_{x-tbc}}\right)^{\gamma}\right]\right\}$$

where $Pt$ is the UE Tx power in dBm, $P_{max}$ is the maximum transmit power in dBm, $R_{min}$ is the minimum power reduction ratio to prevent UEs with good channels transmitting at very low-power level. $R_{min}$ is set by $P_{min}/P_{max}$. $PL$ is the effective path loss in dB for the UE from its serving BS, and $PL_{x\text{-}ile}$ is the x-percentile effective path loss (plus shadowing) value. $PL_{x\text{-}ile}$ is defined here as the value in the CDF, which is greater than the effective path loss of x percent of the MSs in the cell from the BS (i.e. it corresponds to the parameter "power Scale Threshold"). It is set by default to 0.9, but you can change it. With this power control scheme, the 1-x percent of UEs that have a path-loss greater than $PL_{x\text{-}ile}$ will transmit at $\mathbf{P_{max}}$, i.e. are not power controlled. In SEAMCAT, gamma is assumed to equal 1.

*Transmitter settings for OFDMA as interfering link*

Depending on the direction of the interfering OFDMA link to be simulated, it is necessary to pay attention to the emission bandwidth of the unwanted emission mask and the system bandwidth.

When a DL simulation is considered, the unwanted emission mask corresponds to the BS transmitting over all the RBs (i.e. the emission bandwidth is the same as the system bandwidth).

When a UL is considered, the emission bandwidth (i.e. inband part of the unwanted emission mask) corresponds to the UE transmitting over a number of RBs (i.e. the emission bandwidth is equal to the RB bandwidth × number of RBs requested per user), which is different from the DL where the system bandwidth is used as illustrated in

Figure 1.23. Note that the system bandwidth is input to SEAMCAT and approximately RB bandwidth × the total number of RBs (i.e. max subcarriers per BS input).

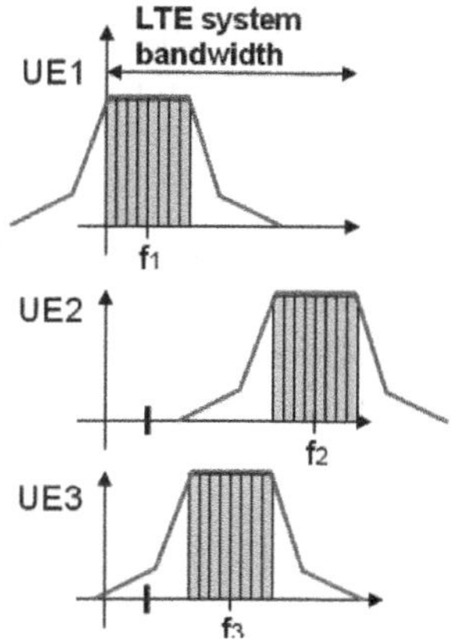


**Figure 1.23.** *Illustration of the emission spectrum mask in UL for an LTE transmission*

| OFDMA DL as interferer | OFDMA UL as interferer |
|---|---|
| When OFDMA is a DL interferer, the OFDMA is not simulated, as it is assumed that the BSs are transmitting at full power and in order to decrease the simulation time, a full OFDMA simulation is not required. In OFDMA DL interferer, the position of the BSs will be calculated only. Figure 1.24 presents the setup of the OFDMA DL as an interferer.<br><br>Note that only the system bandwidth is needed in this configuration, therefore, the rest is shaded (not active). | When OFDMA UL is the interferer, it is important to simulate the whole interfering network (i.e. power control) so that the interfering emission power from the UE is optimized. In this case, the GUI interface is similar to the victim one. |

*Assumptions*

The OFDMA LTE simulation is only valid for a 100% loaded system, and each user is allocated with a fixed number of resource blocks. This is equivalent to modeling a Round Robin scheduler with full buffer traffic model and a frequency reuse of 1/1 (i.e. single frequency network is assumed).

Moreover, E-UTRA system is assumed to be a fully orthogonal system, which indicates that in the UL case only UEs allocated with the same subcarriers (frequency resource block) could introduce other-cell, intrasystem interference.

The number of active users per serving BS is the ratio between the Max subcarriers per Base Station and the Number of subcarriers per mobile (both of these parameters are input). For instance, with 24 RBs at the BS and eight RBs at the UE, the number of active users is three and the system is 100% loaded.

In case where there are 24 RBs per BS and seven RBs, that make three users per BS – but only 21 out of 24 RBs will be in use. Therefore, the system load is equal to $(21/24) \times 100 = 87.5\%$

Note that if the OFDMA is a DL interferer, the OFDMA is simulated as in "traditional" simulation with the BSs transmitting at full power. This decreases the simulation time of a full OFDMA simulation. In OFDMA DL interferer, only the position of the BSs will be calculated because full transmit power is assumed. For all other simulations (including UL), scenarios full OFDMA network simulation is required. Consequently, some of the input parameters of the GUI interface have been gray-out when the OFDMA DL interferer case is selected. Since it is arguable that some simulations assuming a rural environment would not need to assume full power transmission (i.e. full loaded network) when the system is DL and interferer, you may need to manipulate either the input power or spectrum mask (or both) in order to simulate the DL interferer case for rural deployment.

*DL SINR calculation*

In this SEAMCAT OFDMA impThe SINR or C/I calculation in DL is calculated as

$$C/I = \frac{C(j,k)}{I(j,k)}$$

where $C(j,k)$ is the received power at the $k$-$th$ user from the serving BS, i.e. the $j$-$th$ BS

$$C(j,k) = P_{BS}^{UE} \times effective\_pathloss(BS_j, UE_{j,k})$$
$$C(j,k) = dRSS(BS_j, UE_{j,k})$$

and where:

$P\frac{UE}{BS}$ is the power of resource block. Note that the effective path loss includes shadowing;

$I(j,k)$ is the sum of the interference power (power of resource block *effective path loss including shadowing) that consists of adjacent cell interference $I_{inter}(j,k)$ (from the same victim system, i.e. denoted intersystem interference).

$$I(j,k) = I_{inter}(j,k) + I_{ext}(j,k) + N_t$$
$$I_{inter}(j,k) = \sum_{l=1, l\neq j}^{N_{cell}} P_{BS}^{UE} \times effective\_pathloss(BS_l, UE_{j,k})$$

The interference from external interfering system(s) is in adjacent channel $I_{ext}(j,k)$ (interference power into this resource block including adjacent channel interference ratio (ACIR)). The ACIR is implicitly taken into account when both unwanted and blocking mechanisms are summed in the computation

$$I_{ext}(j,k) = \sum_{m=1}^{N_{Kthermal\_cell}} iRSS_{unwanted}(BS_m, UE_{j,k}) \times iRSS_{blocking}(BS_m, UE_{j,k})$$

where:

$$iRSS_{unwanted}(BS_m, UE_{j,k}) = iRSS_{unwanted}\text{(over the size of UE resource blocks)}$$

for each of the UE's frequency where the DL information is received and

$$iRSS_{blocking}(BS_m, UE_{j,k}) = iRSS_{blocking}(\text{over } systembandwidth) \times \frac{N}{M}$$

at the victim system frequency, where N is the number of RBs (i.e. subcarriers) requested per UE, and M is the maximum number of RBs per BS, and where *N_external_cell* is the number of external interfering BSs and the thermal noise $N_t$, where $N$ is the number of RBs scheduled to a UE.

$$N_t = 10\,\hat{}\,((-174 + 10\log 10(\text{bandwith of } N \times RBs) + NoiseFigure_{UE})/10)$$

*UL SINR calculation*

The SINR or C/I calculation in UL is calculated so that *C(j,k)* is the received power from the $UE_{j,k}$ at the *j-th* BS.

$$C(j,k) = P_t(j,k) \times effective\_pathloss(UE_{j,k}, BS_j)$$
$$C(j,k) = dRSS(UE_{j,k}, BS_j)$$

where *Pt* is the transmit power of the UE in dBm (note that UL power control is applied). Similarly to DL, the interference is derived from

$$I(j,k) = I_{inter}(j,k) + I_{ext}(j,k) + N_t$$

where $I_{inter}$ is the interference coming from UEs of the same system but from adjacent cells (i.e. the intersystem interference from other cells). Since a fully orthogonal system is assumed, only UEs that transmit in the same frequency subcarriers will introduce interference to each other, hence only UEs in other cells with the same k index are considered.

$$I_{inter}(j,k) = \sum_{l-1, l \neq j}^{N_{cell}} P_t(l,k) \times effective\_pathloss(UE_{l,k}, BSj)$$

where $I_{ext}$ is the interference from external interfering UEs.

$$I_{ext}(j,k) = \sum_{m-1}^{N_{Kthermal\_cell}} \sum_{v-1}^{K} iRSS_{blocking}(UE_{m,v}, BS_j) \times iRSS_{unwanted}(UE_{m,v}, BS_j)$$

where $K$ is the number of UEs in the external interfering cells and the number of external cells is limited to $N_{External\ cell}$ and the thermal noise $N_t$.

$$N_t = 10 \wedge ((-174 + 10\log 10(\text{bandwith of } N \times RBs) + NoiseFigure_{UE}) / 10)$$

In UL, it is important to remember that for LTE technology, each user will be transmitting its own RB.

When the OFDMA UL is the victim system, we have to remember that the interferer will impair each of the signals transmitted by the UEs serving its own BS (i.e. the victim BS). Therefore, for a specific link (UE1 to BS1) the interference caused by an external interferer will only affect the spectrum occupied by the RBs allocated to UE1 for that link and not the whole system bandwidth at BS1.

*OFDMA LTE link-to-system level mapping*

A look up table is used to map throughput in terms of spectral efficiency (bps per Hz) with respect to calculated SINR (= C/(I+N)) (dB) level. This link level data (bitrate mapping) is user selectable and can be modified depending on the simulation to perform.

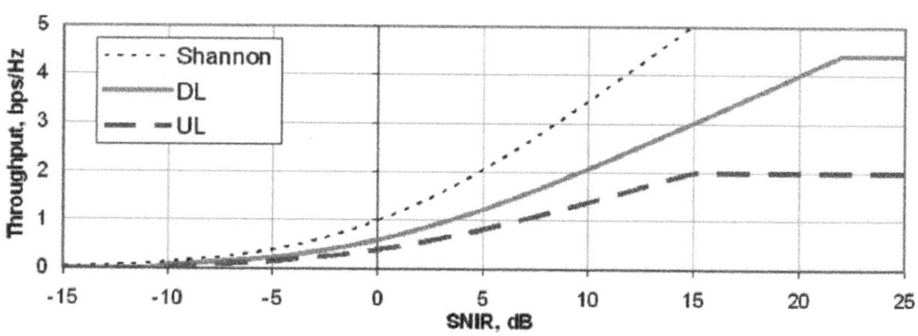


**Figure 1.24.** *SNR*

*Note on the ACLR calculation*

The ACLR calculation is similar to the unwanted calculation BUT note that in 3GPP it is the integration of the interfering power in the adjacent channel where the bandwidth equals to the interfering emission bandwidth while the unwanted uses the victim bandwidth (see illustration from 3GPP TR36.942).

| Parameter | Description |
|---|---|
| OFDMA link component | The type of OFDMA system. There are considerable differences between modeling of uplink and a downlink in OFDMA system |
| SINR minimum | Lower boundary of SINR to take into account in the simulation. In DL, any UE with a C/I lower than the SINR minimum will be disconnected right away. In UL, the UE will get tagged with a disconnect flag. For a specific threshold (maximum allowed disconnection attempts) of disconnection, the UE is removed from the cell |
| Max subcarriers per base station | Number of available resource blocks (RBs) per BS |
| Number of subcarriers per mobile | Number of RBs per UE. Note the ratio of maximum subcarriers per base station/number of subcarriers per mobile gives the number of active users per serving BS |
| Handover margin | Specifies the maximum difference, in dB, between the links in users active list. The actual active list selection is based on path loss calculations |
| Minimum coupling loss (dB) | The minimum path loss. It is used in the calculation of the effective path loss |
| System bandwidth | Specified in MHz |
| Receiver noise figure | Equipment-specific noise figure of receiver, specified in dB |
| Bandwidth of RBs | Specified in MHz |
| OFDMA LTE link-to-system level mapping | The traffic (i.e. bit rate) per UE is a look up table used to map throughput in terms of spectral efficiency (bps per Hz) with respect to calculated SINR (= C/(I+N)) (dB) (signal to interference-plus-noise ratio) level. A drop-down selection of link level data look-up two-dimensions functions from library. The OFDMA link level data has the same formats for uplink and downlink but with different values. This link level data (bitrate mapping) is user selectable and can be modified depending on the simulation to perform. You are responsible to choose an appropriate set of data |

**Table 1.6.** *Parameters*

*OFDMA general settings*

These are similar for the OFDMA DL and UL. The achieved bit rate is calculated as follows:

$$BiteRate_{[kbps]} = \frac{N_{Subcarriers-per-UE}}{N_{total\_subcarriers}} \times \left(x_{bps/Hz}\right)_{SINR} \times BW_{[MHz]} \times bps\_to\_kbps\_conversation$$

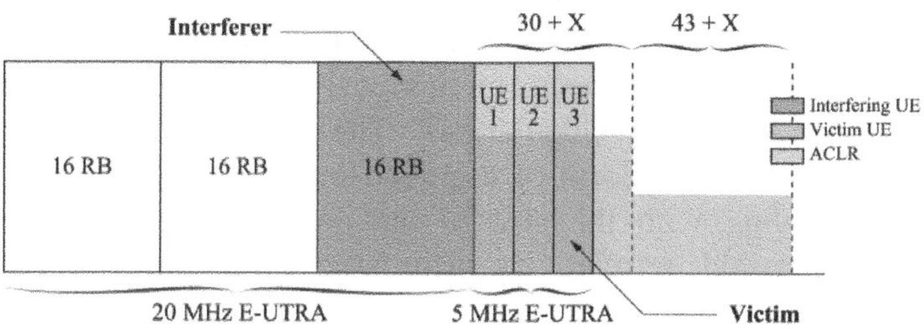


**Figure 1.25.** *ACLR for a 20 MHz eUTRA UE aggressor to a MHz eUTRA UE victims (TR 36.942)*

## 1.5. Radio propagation software

Most mobile operators have developed their own prediction software. Academic experts have also worked on the possible algorithms. Since propagation is essentially a probabilistic phenomenon, it is no surprise that mathematicians have been interested in elaborating sophisticated algorithms in order to fit the reality of measurements. One of the most well-known techniques coming from academia is "optimization by simulated annealing" initially proposed in [KIR 83].

Since the purpose of propagation calculations is to determine a path loss, either to optimize the coverage of a radiocommunication (or broadcasting) network, or to master interferences, the results are in decibels. Therefore, engineers have long worked on heuristic models, which are calibrated by myriad measurements. Among them, let us quote Millington, Deygout, Epstein and Peterson. In a previous book, a comparison is made on the results of these approaches.

Nevertheless, today the leading algorithm for such calculation is the Hata propagation prediction and its derivatives, all COST 231. The calculation is easy. It is based on the well-known approach of terrain cuts: from the transmitter to the receiver, the terrain is cut along the vertical plane linking the two locations. This process enlightens the possible diffractions or masks with their impact, which is calculated by the above-mentioned algorithms, but also makes it possible to add the different contributions of the ground clutter and constitution (sea, lake, forestry, swamps, deserts, etc.). The loss brought by all kinds of ground occupancy has been meticulously measured and is now available for computer calculations. The computer is fed with a digitized terrain in the form of digitized maps, the price of which is inversely varying with the size of the quantum: 200 m × 200 m quantization would be relatively affordable while 1 m × 1 m, when available, will be very expensive. On this quantum, the geographers have averaged all the data: the average height on the quantum will not show that there is some peak or some deep hole in the middle of the quantum.

With the availability of very powerful computers, it has been possible to just try and mimic the actual propagation of the radioelectric waves: this approach is called ray tracing. It is particularly useful for the engineering of microcells, the height of which is low (around 10 m). The issues with ray tracing are not the algorithm as a first cause of errors but:

– error digitized maps: the detailed map of “La Defense” business district of Paris is always late and does not show the latest buildings;

– error location of transmitters. Ray tracing needs that transmitters and receivers be located with an accuracy of a few centimeters. Therefore, the operator has to secure the x, y and z positions with differential GPS.

On a particular terrain cut, it is possible to apply several algorithms the one after the other, e.g. ray tracing for the first hundreds of meters, then Deygout model, then COST 231, etc. The result is just the addition in decibels of all these contributions.

The use of computers for propagation calculation makes it possible to automatize the whole process. The computer can find in the digitized map the possible cell sites that fit the honeycomb grid in force on the studied area, calculate their coverage as well as their possible harmful effects. It is possible to enrich the calculation with the possibility to find electric power in a chosen location and other environmental constraints.

To conclude with this short presentation, let us focus on the presentation issues. The problem with propagation prediction software is that it has to be interfaced with the computer OS. Software changes every 3–5 years, so there is a major effort in providing a stable environment where the sophisticated algorithms will be continuously efficient and provide the same man–machine interface showing coverage maps and data results. Quite a few companies offer such environments (Planet, ETDI, etc.). Generally, these companies have their home calculation product, based on the known algorithms, but they offer a way to integrate the efficient calculation core of the operator.

## 1.6. Macrocells, microcells and femtocells

### 1.6.1. *Macrocells*

As the capacity and coverage requirements of LTE mobile networks increase, service providers require fast and cost-effective solutions to remain competitive. They need an LTE eNodeB that is scalable, high-capacity base stations, providing a full network deployment solution within a single unit.

The required eNodeB should provide integrated packaging of internal antennas (as well as connections to external high-gain antenna components), RF and baseband. The eNodeB will have an internal GPS module with external GPS antenna. This architecture combined with rapid, easy installation and roll-out makes the base station an ideal solution for operators seeking LTE deployment with reduced capital expenditure and maximum return on their network deployment.

In order to easily accommodate the increased complexities inherent in fourth-generation LTE technologies, the LTE eNodeB should utilize software-defined radio coupled with highly integrated systems-on-a-chip (SOC). The base station should combine multiple, task-oriented processing engines in a single scalable, power-efficient solution.

LTE eNodeB features include transmission power of 2 × 1 W and two receiver chains, on a 2 × 2 DL MIMO and collaborative MIMO of 2 × 2 in UL.

### 1.6.2. *Femtocells*

Femtocells are low-power wireless access points – originally called access point base stations – that operate in a licensed spectrum to connect standard mobile devices to a mobile operator's network using the customer's DSL or cable broadband connection.

A femtocell is a scalable, multichannel, two-way communication device that incorporates key elements of a mobile radio access network into a compact device – about the size of a typical desktop Wi-Fi router – and can be deployed in a home or an office.

Femtocells may also be the means of covering an urban area with novel engineering. Like the coverage that is built in some metropolises, such as San Francisco for example, the coverage is ensured by thousands of discrete devices spread all over the streets and populated areas (stores, malls, public buildings, etc.)

The femtocell access point is connected to a high-speed Internet or other IP connection to interface with the core packet-switched networks of LTE. Femtocells work with standard devices that are compliant with the air interface technologies. This ensures seamless service and good interoperability with existing networks and avoids the need for specifically adapted handsets.

Some capacity offload is anticipated from the use of femtocells but these benefits have not yet been quantified.

The femtocell solution typically employs power and backhaul via the user's existing resources (for example, DSL). It also enables a capacity equivalent to a full 4G network sector at very low transmit powers, dramatically increasing battery life of existing phones, without needing to introduce Wi-Fi-enabled handsets.

Potential challenges to the deployment of femtocells are the reliance on the consumer to support the backhaul capabilities. With the advent of fiber to the home (FTTH), this issue should be solved. There is another issue with the possibility of interference caused by the close placement of multiple femtocell devices.

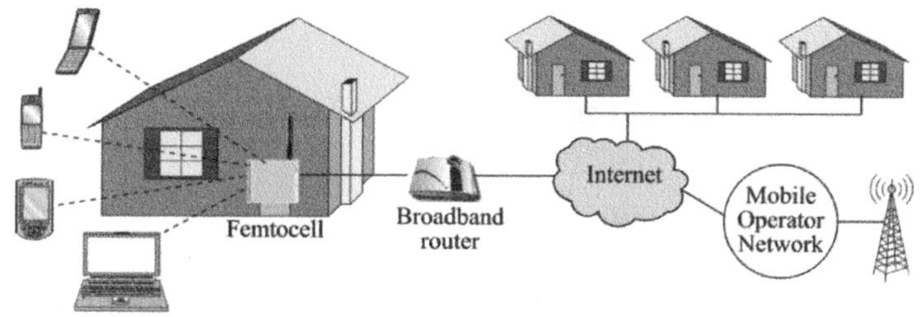


**Figure 1.26.** *Femtocell at home*

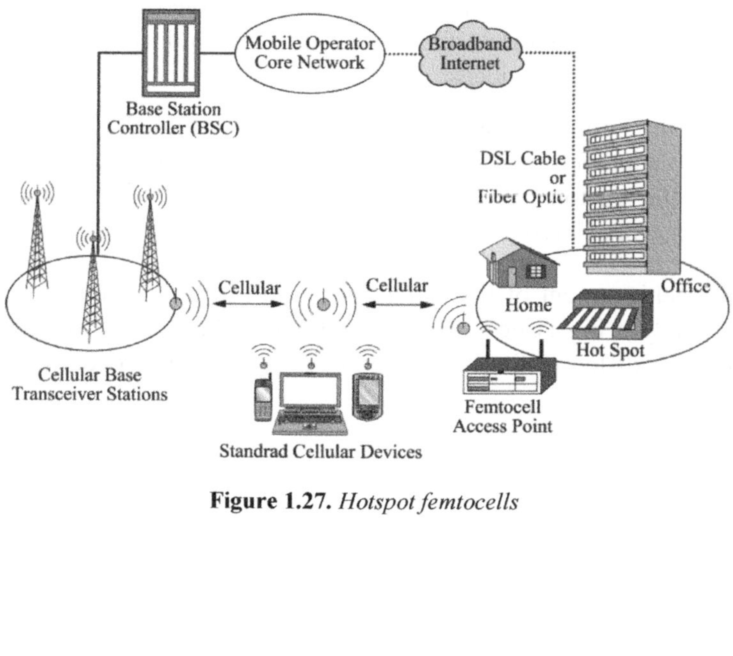


**Figure 1.27.** *Hotspot femtocells*

### 1.6.3. *Remote radio heads*

It is possible to extend the coverage area using black optical fiber to transport the uplink and downlink radio channels in a more appropriate location.

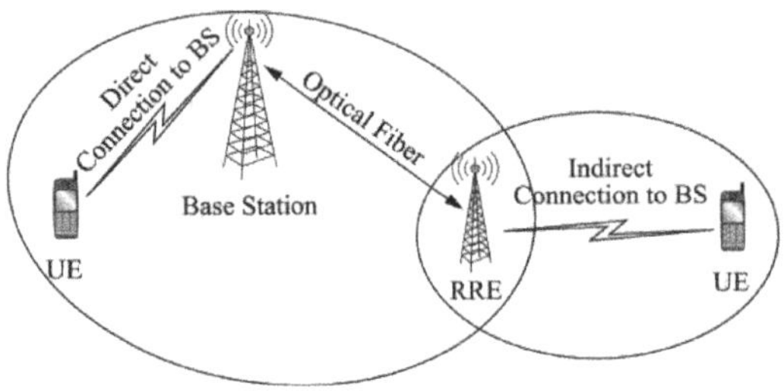


**Figure 1.28.** *Offcentering the radio head*

### 1.6.4. *Heterogeneous network*

The management of a very different kind of cell in the same network creates interesting issues for the servicing of mobiles. The main principles are:

– serving the mobile with the smallest available cell. That cell has been installed to capture a maximum of traffic and offload the upper layers, especially macrocells;

– but the network must have the capability to quickly transfer the mobile to another small cell or to a bigger one (or another small cell) without the customer being hindered, seeing a drop in quality of service or worse having his or her call cut.

The achievement of the signaling plan in heterogeneous networks was already a big issue with GSM networks, having to optimize the use of two different frequency bands as well as macrocell/microcell management.

## 1.7. Backhaul

Generally, the center of interest in a mobile network is the air interface and the deployment of radio network elements. Of course it is important, but a big part of the main intelligence is located in the core network. More forgotten is the backhaul, i.e. the enormous investment in transmission links connecting all the base stations to the intelligent network elements. Thus, the optimization of the backhaul is a key question for engineers of mobile operators. More than often, this backhaul will connect not only LTE elements but also previous generation devices.

### 1.7.1. *The unified backhaul*

The unified backhaul and core network in a hybrid environment of 2G, 2.5G, 3G and 4G systems are depicted in the chart below. The backhaul access/ preaggregation/aggregation uses the options described earlier. For example, the legacy 2G and 2.5G systems use PDH Microwave, SDH and IP-MPLS in the access/preaggregation/aggregation part of the backhaul. The NodeB in 3G UMTS uses either Packet-based Microwave, EoSDH and IP-MPLS option or GPON-Metro Ethernet and IP-MPLS option in the access/preaggregation/aggregation part of the backhaul.

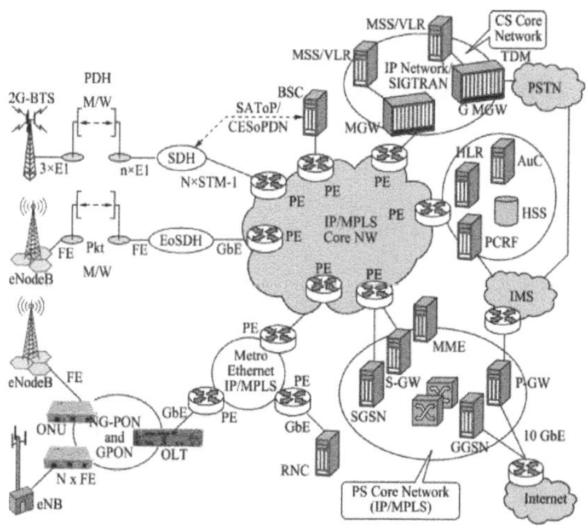


**Figure 1.29.** *Unified backhaul and core network in 2G/2.5G/3G/4G networks*

The nation-wide IP-MPLS network plays an important role in linking all the component networks of a wireless network comprising

2G/2.5G/3G /4G systems. The user data traffic, voice traffic, signaling and OAM traffic, 4G network user data/signaling/OAM traffic, IMS traffic are transported through nationwide IP MPLS network. The traditional TDM traffic of 2G/2.5G is carried using structure-agnostic TDM over packet (SAToP) or circuit emulation service over packet switched network (CESoPSN) over nation-wide IP MPLS network if BSCs are not colocated near BTSs in the same city.

### 1.7.2. *Future of Ethernet backhaul*

The future of Ethernet backhaul seems to be very bright. The current highest speed of Ethernet has already reached 10 gigabits/s and it has been very widely deployed across the globe. The next highest speed interface in the Ethernet family is 40G/100G that is almost finalized in the standardization/testing, which is far more sufficient for future generation mobile networks beyond 4G. Before 40G/100G interfaces are widely deployed in the operators network, standardization for the next highest interface of 400 gigabits/s has already started. This confirms that Ethernet backhaul is going to stay for some time longer as a predominant backhaul technology that can support 100 s of gigabits/s data traffic, which will improve the customer's quality of experience (QoE).

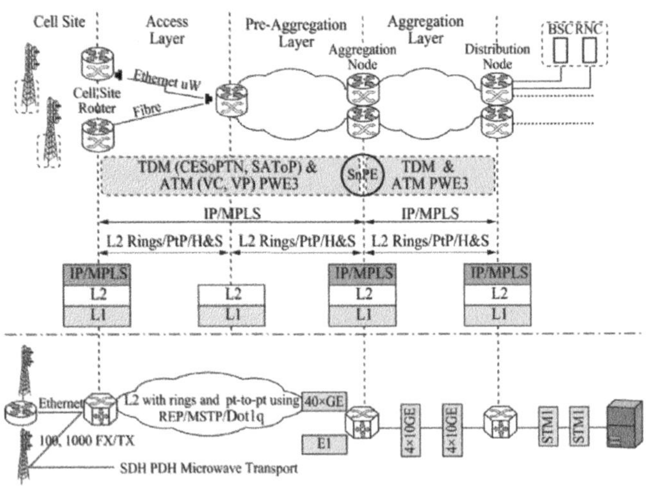


**Figure 1.30.** *Architecture*

In Figure 1.30, there is an assumed intermediate layer 2 POP, and there are rings in the preaggregation part of the network. If we consider the Cisco Resilient Ethernet Protocol topology control in the preaggregation layer, the MPLS/IP RAN built over Ethernet-bridged infrastructures (physical fiber rings or microwave rings) can rely solely on the layer 2 topology protection. This means that layer 2 protection scheme will automatically provide protection for the label switched path (LSP) and ATM/TDM pseudowires. As we are relying on layer 2 convergence techniques, we can build the MPLS/IP RAN on static routes between cell site routers and the aggregation nodes. Static routes are only required between the cell site routers and aggregation nodes to enable the MPLS LSPs and PWE3 segments.

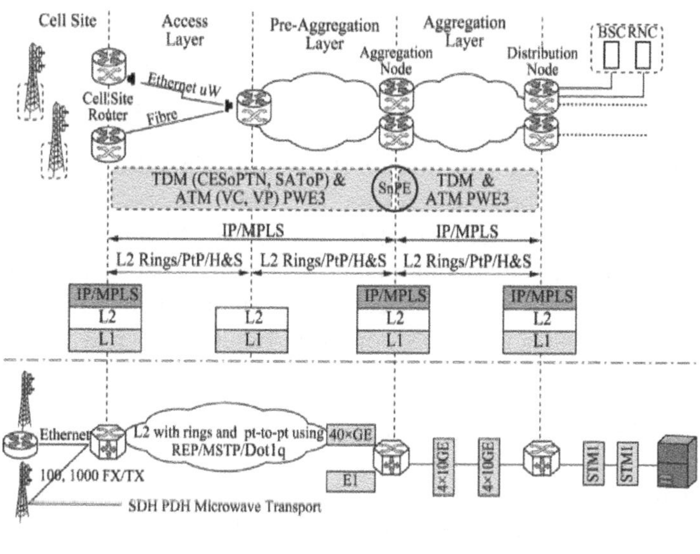


**Figure 1.31.** *TDM and ATM PWE3 backhaul with layer 2 interworking*

Some service providers have little experience of layer 2 Ethernet technologies or believe that there is increased operational complexity caused by the layer 2 control protocols or, more specifically, its integration with MPLS/IP in the aggregation nodes. For such providers, MPLS IP RAN redundancy can rely on IGP/LDP or MPLS traffic engineering (see Figure 1.2). The MPLS/IP RAN IGP (for example, OSPF from the aggregation router to the cell site router) can be configured with bidirectional forwarding detection (BFD) support to provide end-to-end failure recovery in this scenario. In Europe also, operators want to support native MPLS switching on all POPs from the CSG to the aggregation router as seen in Figure 1.2. This architecture allows a common convergence mechanism from end-to-end, including pseudowire redundancy, MPLS fast reroute (FRR), IGP and LDP fast convergence. Many mobile service providers prefer this design, because they have enormous knowledge of MPLS and layer 3 deployments and have not implemented or required layer 2 Ethernet technologies as seen in wireline environments today.

The networks can also be segmented in terms of the IGP/LDP domains by using the switching provider edge (S-PE) capability on the aggregation nodes. In essence, this technique implements multisegment pseudowires (MS-PWs). This design allows static routing in the RAN access for simplicity, while using the dynamic IGP capabilities in the core MPLS/IP domain. This design also allows different IGPs to be used across the radio access and MPLS/IP networks, permitting better overall scalability. In addition, dynamic IGP helps in failure segmentation and isolation, especially considering several RAN infrastructure aspects that can result in IGP instability with technologies such as xDSL or TDM/Ethernet microwave.

### 1.7.3. *UMTS IP NodeB transport over converged packet network*

The initial UMTS NodeBs made use of legacy ATM interfaces only. The initial evolution toward the all-IP vision started with NodeBs supporting an onboard pseudowire capability. This technique did not gather much attraction in the European marketplace, and radio vendors are currently not pushing this solution. There is also support for a hybrid mode on the NodeBs, where the HSDPA traffic is offloaded through an IP/Ethernet interface, and the remaining traffic traverses the ATM interface. Over the last 12–18 months, some prominent European radio vendors have deployed IP NodeBs and IP RNCs with all traffic traversing the Ethernet interface only. Importantly, each NodeB is an IP host (statically configured IP address). The relationship with RNC is still very much a connection-oriented and one-to-one relationship. This evolution is completely different from a wireline environment where the IP digital subscriber line access multiplexers (DSLAMs) and multiservice access nodes (MSANs) are layer 2 devices and switch layer 2 packets. In the European market, Ethernet microware deployment seems to accompany IP NodeB deployments and provides a point-to-point Ethernet (or optionally, ring) access network.

Initial discussions and testing have indicated that there are no plans to support dynamic routing protocols or MPLS on the IP NodeBs in the short-to-medium term. Limitations on the number of IP addresses and static routes supported should improve in later releases. Support for triggering mechanisms, like BFD, is challenging and needed for end-to-end resiliency. There is little support for Ethernet Operations, Administration, and Maintenance (EOAM) capability (connectivity fault management [CFM], 802.3ah, and Y.1731) for fault isolation.

In relation to deployment options that have been used for IP NodeB, there are two main technical operating models considered. The first model is based on layer 2 VPN technologies and could be either an E-Line (point-to-point), E-tree (point-to-multipoint) or E-LAN (multipoint-to-multipoint) service. The second model uses layer 3 or MPLS VPN. We will discuss both models in more detail in the next section.

#### 1.7.3.1. *Layer 2 VPN deployment model*

Initially, European operators often used layer 2 VPNs for connectivity between the IP NodeB and RNC. The connection between the eNodeB and RNC acted as a point-to-point connection, and an appropriate solution was a simple Ethernet pseudowire (E-line). As stated previously, NodeBs tested so far have only been able to support a single static route or single default gateway. The static route in this case points to the distribution layer (see Figure 1.3). Because this is a centralized solution, it is important that there are redundancy options at the distribution layer.

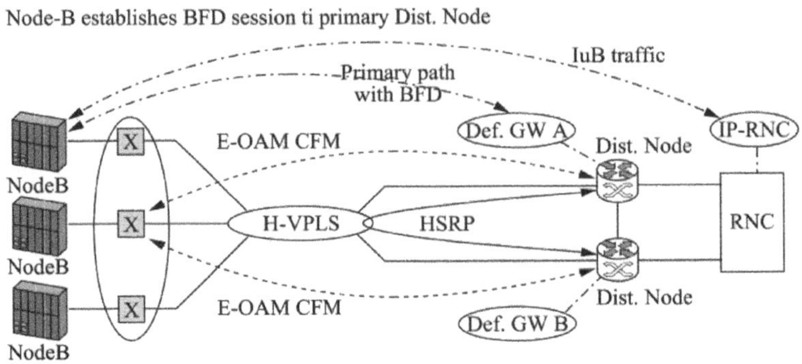


**Figure 1.32.** *IP NodeB layer 2 VPN deployment option*

Currently, hot standby router protocol (HSRP) and virtual router redundancy protocol (VRRP) seem preferable, offering node redundancy with the static route configured on the IP NodeB using the HSRP/VRRP virtual IP address. This solution mandates that an E-tree or E-LAN service (virtual private LAN service [VPLS] or hierarchical virtual private LAN service [H-VPLS]) is required, because a layer 2 path must exist between the distribution nodes in order for HSRP/VRRP to function correctly. Another option is providing a single E-line service from the IP NodeB to each distribution node, but an additional layer 2 path must connect the distribution nodes. This approach addresses resiliency in the uplink direction, but we must consider the downlink direction as well.

In the downlink direction, we must consider end-to-end resiliency. Certain outage types in the layer 2 VPN domain may not be relayed quickly to the distribution nodes, which can result in traffic black holing (lost packets) in the downlink direction. Mechanisms, such as BFD, can help identify a wide range of end-to-end issues and trigger a forwarding change in the distribution node. Ideally, it is best to implement this fast detection mechanism down to the IP NodeB, but this has not been possible in all cases. A trigger for convergence, including some OAM features such as CFM, is a possible answer. Figure 1.3 highlights the resiliency mechanisms currently needed in these environments.

Cisco has seen the use of layer 2 VPNs in operators that have had to use third-party networks for connectivity in the RAN. The third-party network could include an incumbent parent company or in fact, complete outsourcing. In these cases, operators often do not want the third-party supplier to interwork with their routing setup, as would be the case with layer 3 or MPLS VPNs. Instead, these operators often prefer tunneling the traffic with layer 2 VPNs and providing the routing capabilities in their own site to keep overall control. In some deployment, the operator already supports TDM and ATM PWE3 and would prefer to use Ethernet pseudowires for IP NodeB backhaul as well. Ethernet pseudowires will be either implemented in an existing cell site router or on the preaggregation node.

### 1.7.3.2. *Layer 3 MPLS VPN deployment model*

The second option, gaining approval in the last 6–12 months, makes optimal use of the IP NodeB acting as an IP host supporting static routing. The solution, outlined in Figure 1.4, distributes the IP/MPLS capabilities out to the edge of the network. This will allow full dynamic routing capabilities out into the preaggregation and aggregation layers.

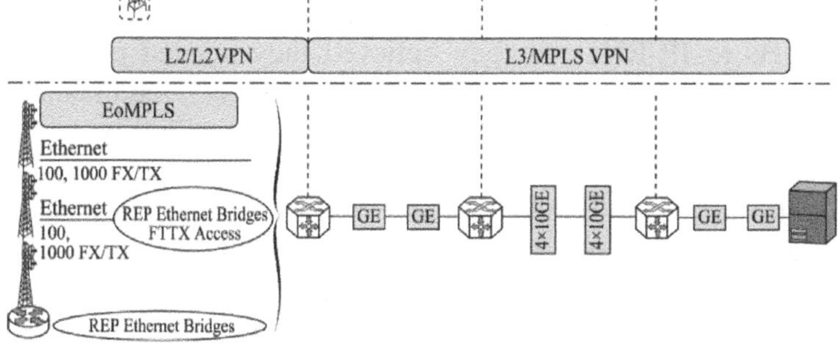


**Figure 1.33.** *IP NodeB layer 3/MPLS VPN deployment option*

This design gives the additional redundancy capabilities of MPLS or IP with similar convergence techniques from end-to-end. This solution also offers the possibility of supporting MPLS VPNs where multiple different traffic types dictate a need for virtualization or isolation. The solution is also an efficient backhaul option with little transport overhead, because there is no emulation. The transport overhead seen with the PWE3 technologies can produce inefficiencies in the order of 200%–300% with small packet sizes. The redundancy options available at the preaggregation edge could include HSRP/VRRP, offering node redundancy. Many European customers employ a single non-redundant circuit in the access (70%–80% of all access circuits). This would be a single point of failure in any case, and therefore node redundancy at the preaggregation level may be unnecessary, making the design easy to provision and monitor.

Also, if a fast convergence triggering mechanism, such as BFD, is required, the design will scale better in distributed environments than in centralized environments. The support of time-based triggering mechanisms, like BFD or CFM, will always present a scaling issue in a centralized environment because of high CPU utilization.

There are some examples in Europe where MPLS is extended to the cell site, normally because a cell site router initiates TDM and ATM PWE3s for GSM and UMTS traffic.

#### 1.7.3.3. *IP NodeB deployment*

Initially, operators were in favor of layer 2 VPN solutions and specifically E-line (point-to-point pseudowires) for connectivity from IP NodeBs to IP RNC, as they believed they needed point-to-point connectivity. However, when matters such as redundancy and scalability were considered, E-tree and E-LAN services were actually required, meaning that the solution was becoming more complex and less controlled than initially thought. Furthermore, for full resiliency end-to-end, a triggering mechanism, such as BFD, was required, and there can be issues with scaling such solutions in large deployments. Overall, the level of complexity has increased in the layer 2 VPN deployments in order to support better resiliency and greater scale. Recent deployments favor using MPLS VPN as far into the RAN as possible, because MPLS VPN offers common convergence and resilience techniques, good virtualization and isolation, and simple integration from access (access links are a single point of failure in most cases) and better scale for triggering mechanisms when compared with centralized implementations.

### 1.7.4. *LTE/EPC transport over converged packet network*

The LTE/EPC evolution is about evolving the radio and core networks toward an all-IP architecture.

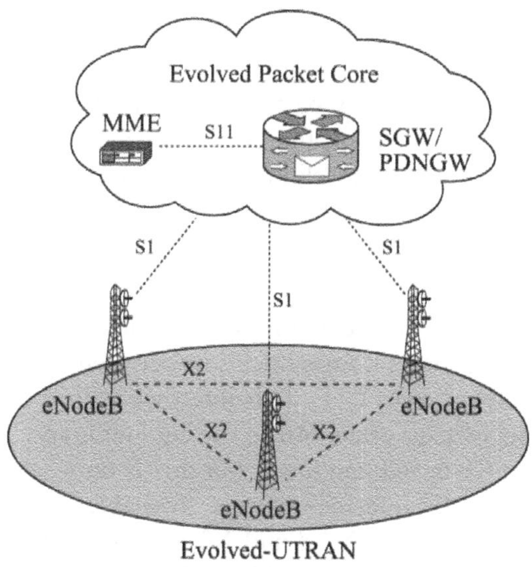


**Figure 1.34.** *LTE/EPC reference architecture*

The radio technology will change from WCDMA to OFDMA, which will result in greater bandwidth and speeds. The flattening of the architecture (removal of the RNC) will result in greater intelligence in the eNodeB. Evolved UMTS terrestrial radio access network (E-UTRAN) is the official 3GPP name for the radio access network of LTE. The X2 interface between eNodeBs will carry control plane (X2-c) and user plane (X2-u) traffic. The core network is now less hierarchical and will contain control plane elements (mobility management entities (MMEs)) with S1 control plane (S1-c) traffic and user plane gateways (serving gateways (SGWs)) with S1 user plane (S1-u) traffic.

#### 1.7.4.1. *LTE factors for consideration with underlying transport network*

The LTE architecture introduces additional requirements on the underlying transport network as highlighted in the following sections.

##### 1.7.4.1.1. Flattened mobile architecture

The traditional mobile infrastructure is very hierarchical with connection-oriented service requirements and one-to-one relationships (that is, IP NodeB has a one-to-one relationship with RNC). The LTE-enhanced NodeB (eNB), now part of the IP infrastructure will have a one-to-many relationship with the core gateways, SGWs and MMEs. This implies that the underlying infrastructure must offer this capability in a scalable and secure manner.

##### 1.7.4.1.2. X2 interface

The X2 interface is a direct communication between the eNodeBs. There was never a direct communication between radio base stations (BTS and NodeB) prior to LTE. This interface will be used for control plane and bursts of user plane traffic during handover. There is also a provision for an S1-based handover, but this is only seen as a fallback option when the X2 interface is not available. Current estimates indicate that the combined X2-c and X2-u traffic could be between 4 and 10% of the core-facing bandwidth (S1-u), and the delay should be less than 30 ms. This traffic is of the utmost importance and from future releases (LTE-Advanced) it is apparent that more user plane traffic will traverse this interface. Also in release 10, there will be stringent latency requirements necessary to implement features such as collaborative MIMO. Figures in the region of 10 ms are currently being considered.

##### 1.7.4.1.3. Distributed architecture

The LTE architecture, compared to other architectures, provides a simpler, less hierarchical model with the capability of simplistically distributing the core gateways. In Europe, there has been much interest in distributing the user plane gateways (SGWs and PGWs) for a number of reasons:

– Bandwidth: some mobile service providers have determined that the bandwidth increases introduced in LTE will massively increase their core bandwidth. In one example, the core bandwidth requirements will increase to 130 Gbps in 2012, based on estimates (current core bandwidth requirements are less than 40 Gbps). In this example, distributing to 12 sites from the previous four core sites avoids upgrading the underlying optical network.

– Traffic offload: some operators are examining the capability of offloading specific traffic types as early as possible in the backhaul infrastructure (also referred as selected IP traffic offload in 3GPP). Operators do not see a value in carrying specific traffic types across core bandwidth. In fact, the operators may be adding little value and so want to hand over the traffic to a third party as soon as possible.

– Video optimization: some operators are carrying large amounts of unicast video, and this accounts for a high percentage of their total traffic, even 70%. The distribution of the gateways allows operators to use technologies, such as caching, offload and local insertion, to save on core transport costs. It is worth noting that the degree of distribution is very important. An example of this would be with very distributed caching that can result in a lower cache hit ratio, and hence requires larger caching capacity.

#### 1.7.4.1.4. Traffic types

There are several types of traffic supported from the eNodeB. Each could have different transport, connectivity and security requirements and will be directed toward different parts of the network. The types of traffic include:

– S1-u traffic destined for the SGW;

– S1-c traffic destined for the MME;

– X2-u and X2-c traffic destined for other eNodeBs;

– operations support system (OSS) traffic destined for core applications that provide fault, configuration and performance management;

– network synchronization traffic.

1.7.4.1.5. Network security and authentication

LTE/EPC is about evolving to an all-IP architecture, and this change provides many advantages in the areas of scalability, availability, flexibility and less hierarchy with direct connection from the radio nodes to the core components. This evolution does introduce some security issues, because now breaches and infiltrations may be possible from the access network. These breaches were never seen in previous mobile architectures and could affect the core gateways directly. For this reason, it is very important a mutual authentication scheme is in place to make sure that the eNodeBs are legitimate and also that the network to which the eNodeB connects is legitimate (hence, mutual). Importantly, the backhaul network is now a carrier Ethernet environment with hundreds or thousands of end users (eNodeBs) who may have varying levels of security. While the network may be private, it is essential to implement all network security features as if building a public network and to choose a transport technology that is most suitable to fulfill this requirement. It is important that the transport technology chosen provides the maximum security possible between eNodeBs. Placing a lot number of eNodeBs in a large L2 domain has already resulted in distributed denial of service (DDoS) attacks. Current investigations explore the capability of extracting the IP address of neighboring cell sites through automatic neighbor relation (ANR) messages for use on dynamic access control lists (ACLs) that will only allow communication between defined neighboring cell sites.

1.7.4.1.6. IPsec requirements

Prior to LTE, end-user traffic would only be decrypted in the core components (RNC for 3G or SGSN for 2G) of the network, which means that all traffic was encrypted when traversing the less secure or third-party networks (unless roaming). In an LTE deployment, the user equipment-to-MME signaling traffic is encrypted. In the 3GPP standard (33.401 section 1.11 and 1.12), there is a requirement to encrypt both signaling and data traffic from the eNodeB (toward the core gateways such as the SGW and MME) when using an untrusted network. However, there is a provision not to provide encryption when the network is considered secure. Similar requirements apply to X2 (control and user). In Europe, an untrusted network is deemed to include such technologies as SDH, PDH or Ethernet Microwave, third-party fiber, hosted or managed last-mile connectivity. This requirement could mean that a security gateway may need to be positioned within the transport network for X2 and S1 traffic. The security gateway concept has led to other areas of discussion including, location of gateways, integrated or standalone gateways, network resiliency options with IPsec, scale and number of IPsec tunnels, key management and IPsec overhead.

#### 1.7.4.1.7. IPv6 requirements

The LTE 3GPP standards contain very detailed information on the support of IPv6 and IPv4 from both host and transport points of view, with a full array of tunneling options as well (IPv4 over IPv6 and IPv6 over IPv4). There is little doubt that IPv6 will become a major design consideration during the lifetime of LTE/EPC deployments. Transitional technologies will need to address the period of time when both IPv4 and IPv6 coexist. A 3GPP study item (TR 23.975) is looking at IPv6 migration guidelines. While the core gateways (PDN gateways) will need to support some of the advanced v6 capabilities (Gateway-initiated dual-stack lite), the underlying network will also need to support both IPv4 and IPv6. There will possibly be a need for carrier-grade network address translation (NAT) capabilities for this transition, and their location in the network will depend on whether a centralized or distributed architecture is deployed.

#### 1.7.4.1.8. QoS requirements

In existing 3G networks, the RAN backhaul presents a challenge for congestion avoidance and the differential treatment of different traffic types or user sessions. The LTE evolution does introduce new concepts, including:

– QoS class identifier (QCI): scalar that controls bearer level QoS treatment; the current specifications have defined nine QCI values (3GPP TS 23.203);

– guaranteed bit rate (GBR): bit rate that a GBR bearer is expected to provide;

– maximum bit rate (MBR): limits the bit rate that a GBR bearer is expected to provide;

– allocation and retention priority (ARP): controls how a bearer establishment or modification request can be accepted when resources are constrained.

Each QCI corresponds to different traffic types (voice, video and so on) and will be categorized with a different resource type (GBR or non-GBR). LTE allows the identification of different traffic types, identification of priority, and the decision about whether to reject the bearer request during resource constraint and then treat traffic in a differential manner. While the LTE standards have made improvements from the previous releases by simplifying the overall QoS mechanism, there are still areas that need addressing, including:

– The standards assume that the underlying network is not contended, which is a major issue with IP/Ethernet deployments. Today's networks are very dynamic, and the available bandwidth is changing (consider adaptive modulation and coding [AMC] with Ethernet Microwave).

– Feedback mechanisms are available to inform the mobile packet core when there is congestion in the radio network. There are no such mechanisms to inform the transport network of issues, and hence packets will continue to be forwarded by the transport network to the eNodeB even under heavy radio congestion. The transport network could prioritize and selectively buffer or drop traffic if there was awareness of the congestion. HSPA cell access control (CAC) includes transport congestion in its mechanism (studied in 25.902 and defined in HSDPA) but this is not defined in LTE.

– Issues occur with the mapping of QCI parameters (nine values) in layer 2 environments where there are insufficient 802.1p bits. While the standards define nine values, most likely more values will be needed for unspecified traffic types (synchronization, OAM and so on).

The underlying transport will need to support traffic prioritization, dual-priority and low-latency queues for 3GPP compliance. Hierarchical QoS (H-QoS) is needed to support the GBR and MBR classification types and also so that important traffic types can be prioritized for multiple different cell sites under congestion conditions. H-QoS is important to manage contention in the last-mile, by representing last-mile available bandwidth at the aggregation and distribution level. Work is ongoing in relation to the bandwidth feedback mechanisms, and protocols such as access node control protocol (ANCP) are under consideration.

1.7.4.1.9. Multicast requirement

Many mobile operators are looking at means to deliver multicast services optimally across their existing networks. Mobile standards have not really addressed this area in a scalable manner. Clearly, other services could use a multicast-type delivery model; these include phone patching, security or software downloads, gaming and so on. LTE and future releases will introduce enhanced Mobile Broadcast Multicast System (eMBMS) with multicast and broadcast modes of operation. Regardless of the modes used, support of source specific mode (SSM) and Internet Group Management Protocol version 3 (IGMPv3), and Multicast Listener Discovery version 2 (MLDv2) snooping on the backhaul network is needed.

1.7.4.1.10. Synchronization requirements

The LTE is primarily concerned with positioning an all-IP solution, and the reliance on legacy networks and infrastructure will be minimal. Capabilities, such as SyncE (Synchronous Ethernet) and packet-based capabilities such as IEEE 1588 version 2 and network time protocol (NTP), are supported to provide network synchronization over the existing transport infrastructure. It is important to remember that LTE may introduce stringent parameters, and support for both frequency and phase synchronization may be required. TDD technologies and LTE multimedia broadcast over a single frequency network (MBSFN) are examples of when phase synchronization is required. Only certain protocols, such as IEEE1588 version 2, have the capability to provide phase synchronization.

#### 1.7.4.1.11. Network convergence

The LTE standard makes use of the GPRS tunneling protocol (GTP), along with stream control transmission protocol (SCTP) for user and control plane connectivity between the LTE/EPC node components (eNodeB, MME, SGW and PGW). The standard only mandates end-to-end connectivity checks with variable intervals and has not specified how the overall network will converge in an optimal manner. SCTP has built-in recovery techniques and requires path diversity for switchover at about 700 ms in 3GPP R4 networks. This presents issues when you consider that this protocol needs to be supported at the eNodeB, because there is a high probability that path diversity will not be present. GTP has inherent path management messages and timers (echo request interval/echo response interval), but the intervals are in the order of tens of seconds, which does not allow optimal convergence. The underlying transport network will provide optimal convergence at an IP layer with mechanisms such as VRRP/HSRP, BGP prefix-independent convergence (PIC), MPLS FRR, IGP fast convergence, IGP loop-free alternates (LFAs) and BFD.

#### 1.7.4.1.12. RAN sharing

European mobile operators have acknowledged that reducing the cost per bit in their backhaul is now their primary objective. Recent commentary indicates that a means being considered to achieve this objective is by implementing RAN or E-UTRAN sharing between different operators. In LTE, E UTRAN sharing is an agreement between operators and will be transparent to the user. This multioperator core network (MOCN) configuration as defined in TS 23.251 is supported over the S1-c and S1-u reference points. This implies that an E UTRAN UE needs to be able to discriminate between core network operators available in a shared radio access network. An E UTRAN sharing architecture allows the operators to not only share the radio network elements, but may also share the radio resources themselves. European operators are currently considering the sharing of resources down to the cell site. This implies that the underlying transport must be able to identify, isolate and provide secure backhaul for different operator traffics over a single converged network. Cisco is working with operators on a means to provide dynamic service creation based on multiple different first signs of life (FSOL). An example would be where traffic on a specific VLAN would initiate a radius request toward a authentication, authorization and accounting (AAA) server that would return information to dynamically setup an Ethernet PW toward the core of the network.

1.7.4.1.13. Fault isolation/identification and fast convergence triggering

As stated before, the LTE standard only mandates end-to-end connectivity checks with variable intervals. This does not help with fault isolation, identification and triggering. The underlying network will be responsible for such capabilities, and there have been proposals submitted for per-link and segment checks. The proposals include:

– layer 1, Ethernet: IEEE 802.3ah OAM;

– layer 2, Ethernet: ITU-T Y.1731/IEEE 802.1ag;

– layer 3, IP: IETF BFD (single hop and multihop).

1.7.4.1.14. Latency requirements

Latency is a key requirement of the LTE/EPC architecture with the goal to achieve a 10–20 ms one-way delay that is an improvement when compared to 100–200 ms in release 99 architectures. The high peak rates and short latency of LTE allow real-time applications such as gaming and IPTV. Latency, jitter and delay parameters must be set for specific interfaces (X2 when supporting some advanced features such as collaborative MIMO). It is imperative that the overall design accounts for these factors and does not introduce excessive latency due to encapsulation (IPsec or unnecessary tunneling that results in suboptimal routing).

1.7.4.1.15. Traffic separation and IP addressing models at the eNode

One of the most important considerations is how the eNodeB will present different traffic types to the backhaul network, because this may be of the utmost importance in determining how the traffic is backhauled to the correct destination. The traffic separation options at the eNodeB are as follows:

– there is no VLAN support, and all traffic is forwarded out the same port;

– traffic divides into two VLANs at the eNodeB. The first VLAN is for X2 traffic that runs directly between the eNodeBs. The second VLAN carries all traffic destined for core applications. This traffic would include S1-u interface, S1-c interface, OSS traffic and so on;

– each traffic type is separated and placed into individual VLANs at the eNodeB.

Radio vendors are showing most interest for the second option (two-VLAN support), because this method represents a good compromise with minimal segregation, because it places traffic that is destined for a similar part of the network into the same VLAN. It also means that there will be no scaling issues in relation to IPsec or VLANs (a minimal number of IPsec tunnels). This option does produce, however, a significant issue, because of the difficulty in differentiating and identifying the different traffic types within the same VLAN. This is important when traffic forwarding goes to the correct end device (SGW, MME or OSS server), possibly through different transport types.

The last option requires the separation of traffic types according to VLAN and gives excellent traffic separation and a means of identifying traffic types toward the core. The technique does present an issue when it comes to VLAN scaling in the network, because each eNodeB could require up to five to six VLANs. More importantly, if each traffic type has its own VLAN, each eNodeB must support five or six instances of IPsec. Radio vendors indicate that this could affect the performance of the eNodeB. Also worth consideration is that the security gateway will need to terminate five to six times more IPSec tunnels, which could also affect the scaling of this platform. For these reasons, it is apparent that shared VLANs will need to be considered for further discussion.

If shared VLANs are considered, then we also need to decide how to identify different traffic types. Here are some suggestions about how this could happen:

– Traffic types could be marked by the eNodeB, but this scenario would assume that all traffic belonging to a traffic type would be treated in the same manner. Because we could be traversing some Ethernet domain, then the number of 802.1p bits supported imposes a restriction also.

– Use of the destination IP address to identify traffic can be complex and is prone to security attacks, because a well-known destination IP address can be spoofed.

– Use of the source IP address to identify traffic is also an option, but this would require that each traffic type be given a separate IP address by the eNodeB. This method could lead to complex IP address planning and address exhaustion.

– Use of IPSec tunnels or child associations to identify different traffic types is another option.

Some believe that the use of traffic marking or destination IP addresses on their own may not be sufficient to identify traffic types. Currently, all different traffic types would be assigned a different IP subnet (/30 proposed). Depending on the deployment model, IPSec tunnels or child associations could also be used in the security gateway as a means of identification before forwarding traffic into an MPLS VPN.

#### 1.7.4.2. *Backhaul technology for an LTE-based converged packet network*

To determine the technical merit of each architecture type, there are ongoing discussions with a number of European operators about possible LTE transport models and the different points mentioned in the preceding sections.

In the European market today, pushing the IP/MPLS control plane out into the RAN and choosing the best possible data plane forwarding technique are becoming a popular option. MPLS VPNs seem to be offering quite an advantage over other forwarding techniques such as layer 2 VPNs, but the overall positioning does not rule out the use of layer 2 VPNs when needed. The model described in the next section highlights the fact that most operators may not be able to get MPLS functionality to the cell site; there may be no active equipment on the cell site. The model shows MPLS functionality going as far as the preaggregation with the option of layer 2 (point-to-point or rings)/pseudowire or MPLS transport profile in the access.

Various types of traffic presented from the eNodeB need individual treatment. The model described in the next section represents the most basic traffic profile model from Europe, with just three different traffic types. In some cases, there are up to six different traffic types. Other traffic types that have been considered include synchronization transport, out-of-band management and closed-circuit TV or cell site monitoring.

#### 1.7.4.2.1. Layer 3/MPLS VPN model for LTE/EPC deployments

For the layer 3/MPLS VPN model as outlined in Figure 1.6, the eNodeB traffic is separated into two VLANs, one for core applications and other for X2 traffic. The core application VLAN must be backhauled toward the core nodes. An MPLS VPN (or half-duplex MPLS VPN) can achieve this when extended over the preaggregation and aggregation layers. The Cisco IOS® MPLS VPN Half-Duplex VRF (virtual routing and forwarding) feature may be helpful, because some operators want to use a hub-and-spoke configuration initially for a configuration like their current one, with no local "hair pinning" and simplified VPN provisioning across the infrastructure.

The advantage to this model for core application traffic is the flexibility of the overall architecture, which can be modified with minimal disruption. If operators can easily insert security gateways for either centralized or distributed IPsec support. This design also offers an advantage to other operators who are looking to distribute some of their core gateways (security gateways or SGW, PGW) in later phases. Cisco also uses common resiliency and availability models right through the preaggregation, aggregation and core networks, which help overcome some of the resilience issues seen in the layer 2 VPN deployments, especially the complexity encountered when a layer 2 VPN service must map to a layer 3 service. A number of European operators have also determined that the use of a single technology from end–to-end without interconnection can reduce operating expenses. An MPLS VPN also offers separation of different traffic types and provides flexible interaction with the security framework. Because MPLS VPN is a layer 3 service, layer 3 attributes can identify and forward traffic or apply different services (QoS, security and so on). The model also provides optimal routing between nodes, which is most important between the eNodeBs; the X2 interface requires direct communication. Features, such as collaborative MIMO, may place strict latency, jitter and delay characteristics on this interface in later releases. The introduction of tunneling in a hub-and-spoke model will incur suboptimal routing and will introduce unnecessary latency (this is also critical when considering the IPsec implementation options).

The X2 traffic is routed through the preaggregation layer using MPLS VPNs. The principle advantage of this method is the optimal routing; in this way, the eNodeBs communicate directly with each other through distributed intelligence. This model optimizes latency and increases bandwidth efficiency when compared with a centralized approach. The MPLS approach provides the ability to control and manage accessibility between the eNodeBs through features such as ACLs, route summarization and so on. Current investigations explore the capability to extract the IP address of neighboring cell sites through ANR messages for use on dynamic ACLs that will only allow communication between defined neighboring cell sites. The MPLS approach will also help in supporting both the direct connectivity model and the model that traverses the IPSec security gateway.

The overall philosophy would be to push the MPLS control plane as far into the RAN as possible and then choose the appropriate data plane for different traffic types. As stated above, this model allows provision of other service types over this converged network when needed. At the bottom of Figure 1.6, we see from the proposed model that a layer 2 VPN could support the transport type related to initial setup, node configuration and software download. A tunneled connection from the cell site into the centralized servers is required, with little interaction with the underlying network and no possibility of breaking out. This method provides a level of security and segregation from other traffic types that are classified as more trusted.

Some operators perceive that pushing IP/MPLS and specifically MPLS VPN capabilities further into the RAN increases the complexity from a configuration and operating expense point of view. The capital expenditure for platforms supporting MPLS VPN would historically have been higher, but more low-end router and switching platforms are now supporting MPLS natively.

1.7.4.2.2. Layer 2 VPN model for LTE/EPC deployments

Using L2VPN technology only for backhauling LTE traffic is a possibility, as outlined in Figure 1.7. The eNodeB traffic is separated into two VLANs, one for core applications and the other for X2 traffic. The core application VLAN needs to be backhauled towards the core nodes in a point-to-point fashion. An E-Line service (Ethernet pseudowire) that can be extended over the pre-aggregation and aggregation layers achieves this backhaul.

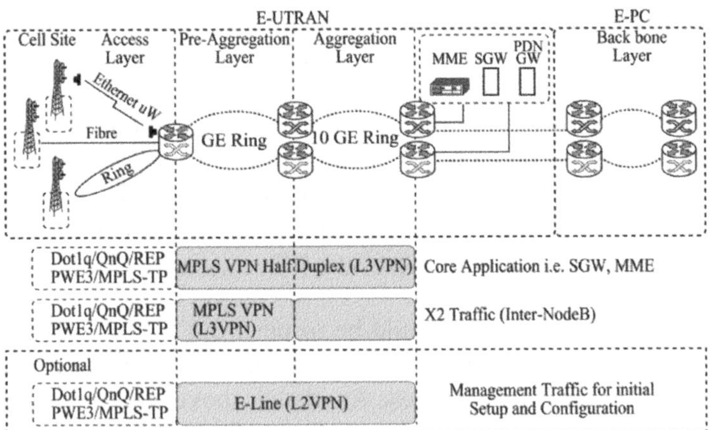


**Figure 1.35.** *LTE/EPC layer 2 VPN connectivity operating modes*

The X2 VLAN will make use of the E-LAN service (VPLS), as the eNodeBs must communicate directly for cell site handover. A critical component of a handover is that the Source eNodeB is able to communicate directly with the Target eNodeB.

While this model presents a very simplistic approach, here are some considerations:

– Supporting the X2 interface by means of an E-LAN service presents an issue, because a mobile user (user equipment) will hand over between different cell sites that must communicate directly with each other. Even if the number of neighbours is low (10 to 15), the issue is that the neighbouring list will change continuously as the user equipment moves from cell to cell. There are two factors that need to be considered: first, the E-LAN domain cannot be so large that it represents a large broadcast domain and hence a security risk; second, different E-LAN domains must communicate with each other to allow handover. Some degree of X2 zoning could be done by connecting the access E-LAN services to pre-aggregation E-LAN services (Hierarchy of E-LAN services). This zoning should be constructed in such a way that cell sites are reachable whenever a cell site handover is possible.

– Using E-LAN services can result in large broadcast domains that present a major security risk, because all eNodeBs in the E-LAN domain could undergo a Distributed Denial of Service (DDOS) breach. Secondly, although the eNodeBs are present in the same E-LAN domain, we only want neighbours to communicate with each other. This segregation in an E-LAN is very difficult to realise and can only be done on a MAC layer through MAC address control access lists, which are operationally complex and not dynamic.

– There could be issues with E-LAN configuration complexity and scaling, because multiple E-LAN services must connect in the access to a hierarchy of E-LAN services in the pre-aggregation layer, to allow handover between cell sites.

– As with deployments seen today involving IP NodeB and IP RNCs, the end-to-end resiliency can present scaling issues and complexity when traversing an underlying E-LAN, E-tree and

E-Line service. Current investigations involve some OAM protocols and mechanisms such as BFD, but there are still some unresolved scaling issues.

– Early analysis conducted on IPSec gateway placement indicated that these gateways might need to be in the transport network in the pre-aggregation or aggregation locations. IPsec termination requires a Layer 3 presence, and this would have implications on any Layer 2 VPN implementation.

– Early indications favor a more distributed approach for security gateway and PDN gateway placement in the later phases. In Europe, an operator is moving from 4 to 6 centralized sites in Phase 1 to 16 to 20 more distributed sites in Phase 2. This distribution is based solely on bandwidth requirements and an issue around the scaling of the underlying optical network. This architecture allows the operator to adopt any offload solution that has currently been analyzed. This approach would have serious effects on the way the Layer 2 VPN model can work and would result in a major redesign of the underlying transport network.

– Some proposed eNodeB authentication mechanisms, such as 802.1x, would have some issues with Layer 2 environments and will not function if there are multiple Layer 2 hops and bridge domains present within the backhaul network.

– Some proposals that promote the use of E-line services, resulting in connection-oriented and centralized backhaul models, will suffer from suboptimal routing and also the insertion of unnecessary latency, which could affect the performance of some features, such as collaborative MIMO or VoIP, in future releases. It also breaks the requirements of having an any-to-any relationship between the radio nodes and the core nodes as outlined in the 3GPP standards.

– A single VLAN with multiple traffic types will present issues when using the Layer 2 VPN backhaul model, as this service will not be able to interpret any Layer 3 attributes. The core would need to support some routing capability to allow transport towards the correct end devices that will be in different IP address subnets.

#### 1.7.4.2.3. LTE/EPC transport conclusions

The LTE/EPC evolution is an evolution towards an all-IP architecture and will fundamentally change how mobile backhaul networks are built in the future. The availability of Ethernet-enabled NodeBs and the evolution towards LTE/EPC pushes IP awareness further into the edge of the mobile network. Mobile operators are beginning to view these backhaul networks like carrier Ethernet environments offering multiple concurrent services. LTE/EPC will make demands on the underlying transport in areas such as security, IPv6, distributed intelligence, multicast, synchronization, QoS, fast convergence, instrumentation, and management. The transport technology choices of today will be important for the future evolution of the mobile architecture. The LTE/EPC evolution demands a lot of intelligence and flexibility in the underlying network. Cisco recommends a design model to support a distributed, multi-service, MPLS enabled network that offers the flexibility, scalability and intelligence to address current and future needs. This design allows the use of intelligent Layer 3/MPLS VPN technology for optimal routing, security, flexibility and resiliency and also provides possible support of Layer 2 VPN technologies if deemed necessary for certain traffic types.

### 1.7.4.3. *Conclusion*

European mobile providers are currently experiencing large increases in mobile backhaul capacity to address their current and future service requirements. The costs and expenditures associated with providing this increasing bandwidth has not being linearly matched by revenue growth. The primary objective is to increase the bandwidth while simultaneously reducing the cost per bit. Existing TDM/ATM infrastructure will neither scale to the required bandwidth nor meet the cost reduction requirement. Recent reports have shown that all operators now believe that IP/Ethernet-based backhaul is a mandatory requirement. These reports also show the growing belief that a single, converged, all-IP-based Ethernet backhaul is required, with 85% of respondents seeing LTE as a key driver for IP/Ethernet-based backhaul.

While there is a clear effort towards supporting IP/Ethernet backhaul, there are ATM/TDM-based requirements for GSM and 3G that need support. Cisco believes that a converged architecture is essential where the mobile backhaul solution simultaneously supports ATM/TDM and Ethernet requirements. The ATM/TDM requirements can be met through pseudowire technology (PWE3), and the current Ethernet requirements can be supported by means of Layer 2/Layer 2 VPN or Layer 3/MPLS VPN technologies. The transport solution chosen for the current Ethernet requirements must allow for future scaling, simplistic and optimal resiliency, and optimal support for future technology such as LTE. Current Layer 2 VPN-based deployments for 3G-based IP NodeBs are showing issues regarding scale and optimal resiliency. A more distributed Layer 3/MPLS VPN approach is showing better resiliency and scale and better support for the service requirements of the evolving mobile standards.

The LTE/EPC evolution is an evolution towards an all-IP architecture and is seen as one of the most important incentives for the adoption of IP/Ethernet in the backhaul. The LTE/EPC evolution will push more intelligence further out into the RAN and onto the eNodeBs with direct interfaces (X2), and requires an any-to-any relationship between the radio and core nodes. These changes make demands on the underlying transport in areas such as security, IPv6, distributed intelligence, multicast, synchronization, QoS, fast convergence, instrumentation and management. Cisco recommends a design model that supports a distributed, multiservice and MPLS-enabled network. This design allows the use of intelligent Layer 3/MPLS VPN technology for optimal routing, security, flexibility and resiliency, but also provides the possibility of supporting Layer 2 VPN technologies if deemed necessary for certain traffic types.

## 1.8. Frequency planning

With LTE, the network is basically SFN (single frequency network). There is no frequency planning the way it was necessary to work on with GSM. LTE just broadcasts on the entire frequency allocation with a wide band signal. Nevertheless this signal is the concatenation of the different subcarriers which carry the information data. The planning of subcarriers will have to follow the same rules as the planning for frequencies in GSM.

## 1.9. Compatibility with DTT

Under 1 GHz, the major part of the spectrum has been for a long time devoted to terrestrial television. In Europe, this pre-eminence of terrestrial television lasted up to 2010–2012, when for the first time a few MHz were pulled away from broadcasters and granted to mobile operators (at a very high cost). This situation shows the political weight of broadcasting, especially in Europe, where terrestrial broadcasting still keeps frequency blocks which are the most efficient for mobile communications:

– cable and satellite have conquered an increasing audience, reaching more than 90% in some countries of Northern Europe;

– the technology of cable and satellite has continuously improved, as in the USA, where they dominate the audio-visual market;

– telecommunications operators introduced high-speed Internet, able to transmit digital television as soon as MPEG4/H264 became operational. Before 2020, H265 will again bring a breakthrough in the digital video compression: HDTV will be available with less than 3 Mbps.

Even with the surprising decision of WRC 2012 opening the 700 MHz band to mobile communications, terrestrial television will still be broadcast in many European countries. Of course, it seems that the countries where the broadcasting audience is marginal, the spectrum will be managed with a lesser priority to broadcasters.

Terrestrial television had to go digital in order to try to compete with cable and satellite: facing hundreds of TV channels offered by them, the only way was to increase the number of TV channels within a given spectrum was by digitizing the signal. Several technologies were available:

– video and sound compression: MPEG2, MPEG4/H264 and now H265;

– signal encoding for broadcasting based on OFDM: following the satellite standards DVB-S then DVB-S2, the terrestrial broadcasting adopted DVB-T followed by DVB-T2.

The issue with digital terrestrial television is not so much with the downlink transmission. For the downlink, it is quite easy to forecast the effect of the cohabitation between DVB_T (or DVB-T2) and LTE on adjacent frequencies. Frequency planning software calculates the interference patterns easily. Of course, the ideal situation will be co-located transmission if possible.

The main problem, and it will be worse with the advent of mobile communications in the 700 MHz band, is the mobile uplink transmission. As a matter of fact, the mobile will radiate in the frequencies where television sets have been built to receive the broadcast emission, in Europe up to 862 MHz. Both LTE and DVB-T/T2 are maximizing the occupation of the spectrum, which is allowed to them. The mobile emission will be received by the TV set and the result will be a black screen. To avoid this damaging situation, it is necessary to insert at the input plug of the TV set a sharp filter (typically a SAW filter) preventing the set receiving the mobile emission. Quite a few trials have been carried out in Europe, which corroborate this evidence.

## 1.10. Health effects

As soon as mobile operators left the financial difficulties, a very fruitful business flourished: the “research” around the health effects of radiowaves. The effects of such waves have been studied since World War II. As a result, the 5W GSM handheld mobiles were forbidden and the networks had to be engineered for 2W terminals.

Since 1996–2000, depending on the country, quite a few "laboratories" searched for grants and fed the newspapers with "results".

The effects of radiowaves are well known. In the VHF and UHF area, the waves have a frequency, which is too low to cause damage in living cells and the impact is only the "microwave oven effect", thermal increase of the organic tissue. Of course, when some "scientists" place a 450 W klystron close to the head of a rat, the unfortunate animal suffers heavy damage of the same kind as the dog whose owner put it in the microwave oven to dry it after a wash.

As an answer to the people who are uncomfortable with the sight of antennas, there are two facts:

– in quite a few cases, these antennas are not transmitting;

– more realistically, there is the example of the Eiffel Tower in the middle of Paris. For decades, the powerful broadcasting transmitters of this tower created a level of more than 6 V/m on the terrace situated across the river Seine, and where most of the wealthy children of Paris skate day after day: no complaints. Of course, some journalists seriously assert that broadcasting waves are harmless – even considering that they apply similar OFDM modulations.

The situation is now showing considerable concern from the public, *now faced with very many highly visible radio masts* – but with little visualization or understanding of radiowave propagation.

Interestingly, in some parts of California, the mobile operators must ensure that antennas cannot be seen easily. Authorities say that it considerably reduces the claims from the surrounding citizens.

### 1.10.1. *Physical facts*

Energy quanta of radiofrequencies less than 3 GHz are far below the level needed to break chemical bonds in DNA. Oscillations induced in ions are too small to cause effects. Resonant absorption by biological tissue still seems doubtful. Cell polarization is negligible. Current flows across cell membranes are negligible.

Ionizing radiations are:

– X-ray (medical, TV screens) – ionizing effect (deep);

– nuclear (natural / power plants) – ionizing effect, radiation hazard is deeper and risk of cancer);

– gamma ray (radioactive process) – ionizing effect (risk of mutation and cancer).

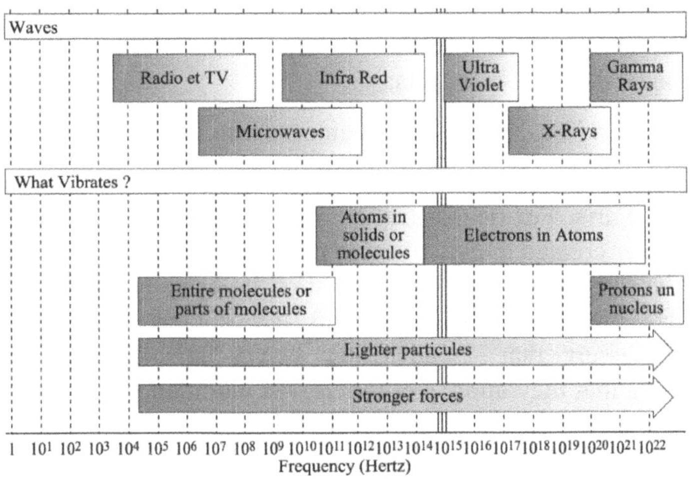


**Figure 1.36.** *Frequencies*

Electromagnetic waves interacting with matter can be reflected, absorbed or transmitted. Microwaves emitted by mobile phone systems are absorbed by human tissue.

What happens depends on the frequency of the electric field and the natural frequencies of the atoms and molecules. The cerebrospinal fluid in the brain is close to water. The absorption curve of liquid water has a resonance at 9 GHz:

– well below the natural frequency: not much absorption;

– near the natural frequency: absorption;

– well above the natural frequency: not much absorption.

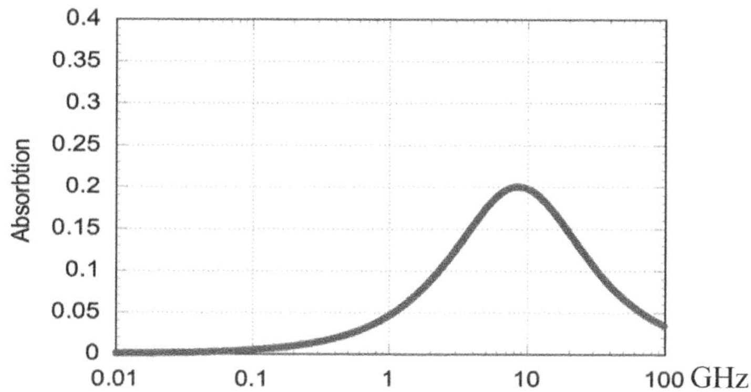


**Figure 1.37.** *Absorption of radio waves by liquid water*

A microwave oven cooks food by *heating it*. The heating comes from intense waves at 2.45 GHz (instead of a wide spectrum of waves at infrared frequencies). Why 2.45 GHz? Microwave ovens operate at 2.45 GHz. This frequency has been chosen to optimize cooking, knowing the absorption properties of water molecules at that frequency:

– if power were absorbed too strongly, microwaves would only penetrate a short distance, the surface would be heated and the inside would remain uncooked.

– if power were absorbed too weakly, microwaves would go right through, without cooking.

– if power is absorbed just right, microwaves penetrate about 5 cm (2 inches), cook the outer 5 cm of the food, which is good enough for most cases.

The *potential hazard* from mobile phones and other wireless devices arises from the *absorption of microwave radiation. Mobile phones* only emit one or two watts, and such a small amount of power makes experiments difficult. *Microwave ovens* emit a few hundred watts and make experiments easy.

### 1.10.2. *Specific energy absorption rate*

The effect of radiowaves on human tissue is measured by the specific energy absorption rate (SAR). It is the basis of the legislation.

SAR can be calculated knowing:

– the electric field level in the organic tissue, $E$, in V/m;

– the density of the electric current, $J$, calculated from the electric and magnetic fields, in Ampere per square meter $\frac{A}{(m^2)}$;

– the density of the organic tissue, $p$, in $\frac{kg}{m^3}$;

– the thermal capacity of the tissue $ci$, in $\frac{J}{kg\ K}$;

– the electric conductivity of the tissue, σ, in $\frac{s}{m}$;

– $K$ is the temperature;

– $\frac{dT}{dt}$ is the derivative of the tissue temperature, in K/s;

$$SAR = \frac{\sigma \vec{E}^2}{\rho}$$

$$SAR = \frac{J^2}{\rho\sigma}$$

$$SAR = c_i \frac{dT}{dt}$$

By law, phones must have a maximum SAR of 2 watts per kilogram averaged over 10 grams of tissue.

1.10.2.1. *SAR: example of SAR calculation using light rather than microwaves*

– *10 cm from 20 W source.* 2 W absorbed in hand. Intensity 200 W per square meter. SAR 20 W per kilogram;

– *1 m from 20 W source.* 0.02 W absorbed in hand. Intensity 5 W per square meter. SAR 0.2 W per kg;

– *1 cm from 20 W source.* Roughly 20 W absorbed in hand. Intensity roughly 8,000 W per square meter. SAR roughly 200 W per kg.

Calculations assume the mass of a hand as 100 g:

– *10 cm from 20 W source.* Roughly 2 W absorbed in hand. Intensity roughly 200 W per square meter. SAR roughly 20 W per kg;

– *1 m from 20 W source.* Roughly 0.02 W absorbed in hand. Intensity roughly 5 W per square meter. SAR roughly 0.2 W per kg.

### 1.10.3. *International Commission on Non-Ionizing Radiation Protection*

The supervision of the issues related with possible health effects of the radio communications is the task of International Commission on Non-Ionizing Radiation Protection (ICNIRP). The tasks are:

– to consider concerns about the possible health effects from the use of mobile phones, base stations and transmitters;

– to conduct rigorous assessment of existing research;

– make recommendations on further work to improve the basis for sound advice;

– identify behavioral changes i.e. a rise in whole body temperature in excess of 1°C at an SAR of 1–4w/kg.

SAR for mobile phones is limited to 2 W/kg, averaged over 10 g. In the USA, the limit is 1.6 W/kg, but averaged over 1G.

ICNIRP basic restrictions on exposure are as follows.

| Tissue region | SAR limit (W/kg) | SAR limit (W/kg) | Average parameters | Average parameters |
|---|---|---|---|---|
| | Occupational exposure | General public exposure | Mass (g) | Time (mn) |
| Whole body | 0.4 | 0.08 | | 6 |
| Head and trunk | 10 | 2 | 10 | 6 |
| Limbs | 20 | 4 | 10 | 6 |

**Table 1.7.** *ICNIRP basic restrictions on exposure*

More specifically for head and trunk:

| Frequency range | Current density for head and trunk (mA/$m^2$) | Current density for head and trunk (mA/ $m^2$) |
|---|---|---|
| | Occupational exposure | General public exposure |
| Up to 1 Hz | 40 | 8 |
| 1–4 Hz | 40/f | 8/f |
| 4 Hz–1 kHz | 10 | 2 |
| 1–100 kHz | f/100 | f/500 |
| 100 kHz–10 MHz | f/100 | f/500 |
| 10 MHz–10 GHz | | |

**Table 1.8.** *ICNIRP basic restrictions on exposure: for head and trunck*

The reference level for general public exposure to time-varying electric and magnetic fields is as follows.

| Frequency range | E field strength (V/m) | H field strength (A/m) | B field strength (μ T) | Equivalent plane wave power density (W/m) |
|---|---|---|---|---|
| Up to 1 Hz | | $3.2 \times 10^4$ | $4 \times 10^4$ | |
| 1–8 Hz | 10,000 | $3.2 \times 10^4 \frac{}{F}$ | $4 \times 10^4$ | |
| 8–25 Hz | 10,000 | 4,000/f | 5,000/f | |
| 25–800 Hz | 250/f | 4/f | 5/f | |
| 0.8–3 kHz | 250/f | 5 | 6.25 | |
| 3–150 kHz | 87 | 5 | 6.25 | |
| 0.15–1 MHz | 87 | 0.73/f | 0.92/f | |
| 1–10 MHz | $87/ f\frac{1}{2}$ | 0.73/f | 0.92/f | |
| 10–400 MHz | 28 | 0.073 | 0.092 | 2 |
| 0.4–2 GHz | $1.375 f\frac{1}{2}$ | $0.0037 f\frac{1}{2}$ | $0.0046 f\frac{1}{2}$ | f/200 |
| 2–300 GHz | 61 | 0.16 | 0.20 | 10 |

**Table 1.9.** *General public exposure*

Recent publications concerning possible health hazards created by radiofrequencies include [LUR 09] and [LUR 11].

### 1.10.4. *Measurements of SAR, experimental studies*

To measure SAR, laboratories now build so-called "phantom heads".

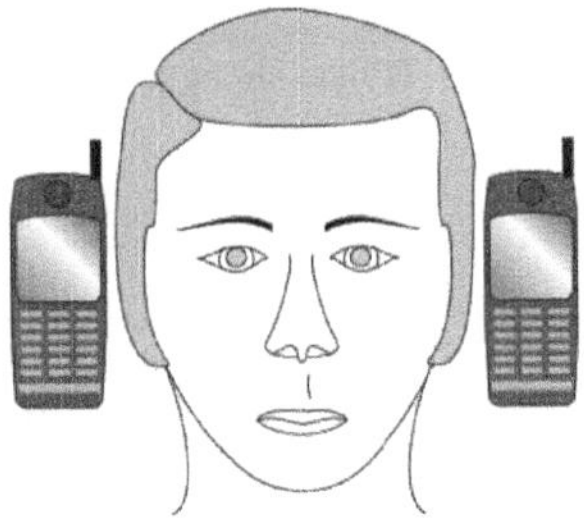

**Figure 1.38.** *Illustration for noise*

Examples of measurements are provided in Figure 1.39.

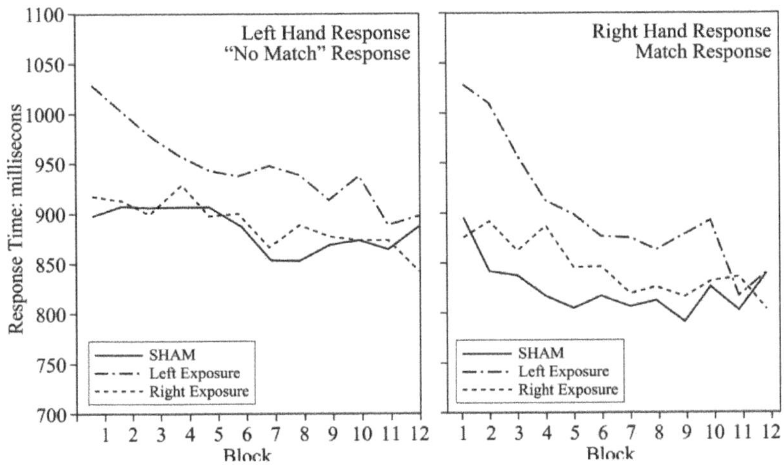


**Figure 1.39.** *Examples of measurements*

1.10.4.1. *Experimental studies – effects of RF radiation on people*

– Nervous system – changes in the brain or behavioural effects;

– cancer-related studies – carcinogenic process;

– effects on the heart and blood pressure;

– brain function:

- studies of the cognitive performance,

- electroencephalogram.

1.10.4.2. *Epidemiological studies*

– People using mobile phones:

- mortality and cancer incidence,

- other health effects;

– exposure to RF radiation through work and hobbies:

- cancer,

- health outcomes other than cancer;

– residence near transmitters.

A massive study, following 1,656 Belgian teenagers for a year, found most of them used their phones after going to bed. It concluded that those who did this once a week were more than three times – *and those who used them more often were more than five times* – as likely to be "very tired" (*The Independent*, 20 January 2008).

### 1.10.5. *Comparison of SAR caused by different devices*

1.10.5.1. *RF sources we encounter daily*

– Broadcast (TV/radio) – kW in VHF/UHF;

– trunk portable phones (5 W in VHF/UHF range);

– pager/cordless phone (< 1 W in VHF);

– microwave oven – source produces 2,000 W, but only 5 mW leaks out of the door (2.45 GHz);

– cellular phones operate in 800/1,900/2,600 MHz bands, cell tower power can be up to 25 W; phone can put out 0.5 W (800, 1,900, 1,700 and 2,100 MHz);

– wireless LAN/WiFi (access points power is <1 W, PDA power is in mW) – 2.4 and 5.3 GHz;

– bluetooth devices; wireless keyboards and mice; DECT cordless phones; baby monitors; “walkie talkies”;

– satellite communications 4–40 GHz; they point toward the sky and do not interfere with the public;

– microwave repeaters 4–80 GHz. These antennae look at each other, do not interfere with the public.

#### 1.10.5.2. *SAR values*

In a microwave oven, 700 W provides 140,000 V/m (Volt per meter). SAR for 1 kg of water will be 700 W/kg.

| | SAR (Watt per kg) | Temperature rise in 1 kg of “brain fluid” |
|---|---|---|
| Microwave oven | 700 | 8°C (ish) |
| Mobile phone | 1 | 0.01°C (ish); cannot be measured directly |

**Table 1.10.** *Microwave SAR versus mobile phone SAR*

Mobile phones are “radio” phones operating at microwave frequencies. Handset power is minimized by having a network of local transmitters and receivers (52,500 base stations in UK).

*Handset* power is less than 1 W. Peak power is 2 W max, average power 0.25 W max. When not engaged in calls, the *handset* sends short signals every few minutes (or few hours, depending on the amplitude of the traveling of the subscriber) to establish which is the nearest mast.

Mobile phone SAR: typical SAR with phone near the head

| Power | Intensity | Maximum SAR |
|---|---|---|
| Watt | Watt per square meter | Watt per kg |
| 1 | 200 | 1 |

**Table 1.11.** *Mobile phone SAR: typical SAR with phone near the head*

For cellular systems, mast power is between 60 and 120 W.

Base station SAR is related to the equivalent isotropic radiated output (EIRP). The typical radiation scheme of a base station shows a narrow beam in the vertical plane.

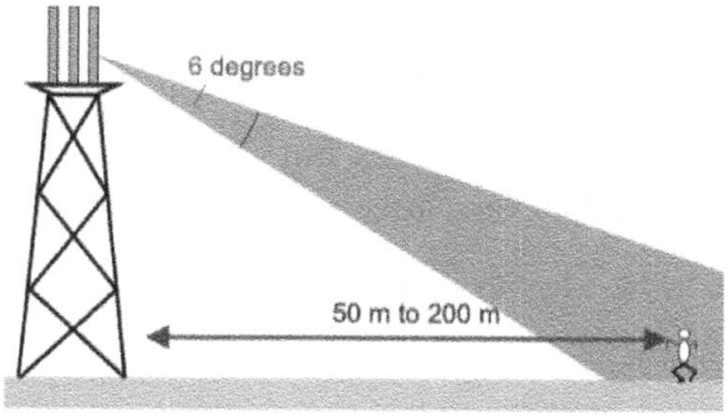


**Figure 1.40.** *Base station narrow beam*

Right underneath the base station: 0.3 W/$m^2$ maximum.

100 m away: 0.01 W/$m^2$ maximum.

| | Power (Watt) | Intensity (Watt per square meter | Maximum SAR (Watt per kilogram) |
|---|---|---|---|
| Handset | 1 | 200 | About 1 |
| Base station | 60 | 0.01 | About 0.001 |

**Table 1.12.** *SAR*

Base station SAR is extremely low!

*Wi-Fi*

A transmitter emits: maximum permissible power in EU EIRP = 0.1 W

1 m from base station, intensity is 0.008 W per square meter.

| Power | Intensity at 1 meter | SAR at 1 meter |
|---|---|---|
| Watt | Watt per square meter | Watt per kilogram |
| 0.1 | Less than 0.01 | About 0.0001 |

**Table 1.13.** *SAR at 1 m*

### 1.10.6. *Safety limits – towers*

1.10.6.1. *Tower types*

– Broadcast communication (TV and radio):

- 10 MW max, 10 kW or less typical;

- Broadcasts are high power, but one-way systems. (TV/radio units do not transmit, they only receive).

– Communication towers (professional communications and trunk):

- 100 W power at antenna but the power reduces exponentially as the sphere expands (similar to dispersion of visible light starting from a light bulb);

- mobile in bus or trains (10 W typical); portable radio (5 W typical).

– Cellular antenna towers/access points:

- Cell tower (25 W, max, 10 W typical);

- Cell phone transmit (0.1 mw to 500 mW).

The engineering of cell towers is based on directive antennas, which have a null at the vertical of the antenna and a narrow beam. The least exposure is just under the antennas.

| Towers | FCC /OSHA | Typical | Comment |
|---|---|---|---|
| Broadcast tower (radio or TV) | 8 W/kg of body mass (below 450 MHz) | 100 KW to 1 MW at the tower | Within safety limit at the either TV/radio receiver (in premises). |
| Cell phone tower – public | 0.08 W/kg over whole body | 10–25 W at the tower | Below 0.08 W /kg for public |
| Comm. tower (professional services) | 8 W/kg of body mass | 100 W at the tower | Below 8 W/kg at portable |

**Table 1.14.** *Example of different towers*

### 1.10.6.2. SAR limitations in the USA

The limits for the terminals, which are next to the body are as follows:

| Devices | FCC /OSHA | Typical | Comment |
|---|---|---|---|
| Portable phone (VHF/UHF) in controlled environment | 7 W/kg in the 300 KHz to 1 GHz range | 5 W at the handset (work related/professional) | 5 W/kg at worker level – constant |
| Cell phone/mobile phone/PDA/scanner | 1.6 W/kg over 1 g of body mass, 4 W near hands, wrists, feet and ankles | 0.1 mW to 0.5 W at the handset | 0.5 W if user is at edge of the cell, 0.1 mW if user is near a cell tower + |

**Table 1.15.** *Safety limits – proximity devices*

## 1.11. Appendix 1: radio dimensioning and planning exercises (courtesy of Emmanuelle Vivier)

Exercise on LTE radio dimensioning and cell capacity.

First, take into account the performance of the UE and the eNodeB as follows:

| Characteristic | Unit | Uplink | Downlink |
|---|---|---|---|
| EIRP | dBm<br>W | 21<br>0.125 | 47<br>50 |
| Rec. Thres. $_{QPSK}$ | dBm<br>W | –126.5<br>$2.2\ 10^{-16}$ | –106.5<br>$2.2.10^{-14}$ |
| PL $_{QPSK}$ | Db | 147.5 ⇒144.5 | 153.5 ⇒150.5 |
| Rec. Thres. $_{16QAM}$ | dBm<br>W | –120<br>$10^{-15}$ | –100<br>$10^{-13}$ |
| PL $_{16QAM}$ | dB | 141 ⇒ 138 | 147 ⇒ 144 |
| Rec. Thres. $_{64QAM}$ | dBm<br>W | –113<br>$5\ 10^{-15}$ | –93<br>$5.10^{-13}$ |
| PL $_{64QAM}$ | dB | 134 ⇒ 131 | 140 ⇒ 137 |

*PL is the maximum path loss accepted by the LTE system as deployed in this case.*

*The uplink is the limiting direction when no traffic is in the cell.*

The propagation model will be the COST 231 extension to the Hata model, which gives a simple logarithmic formula, and a linear equation in dB.

The calculations will be made with:

– a dense urban area;

– 30 m high base stations.

*L is the path loss in these formulas.*

$$L_{dB}(d) = 28.8 + 33.9 \times \log_{10}(d) + 35.22 \times \log_{10}(d)$$

This general formula results in:

– with $f = 2.6\text{GHz}, L_{dB}(d) = 144.7 + 35.22 \times \log_{10}(d)$

– with $f = 1.8\text{GHz}, L_{dB}(d) = 139.2 + 35.22 \times \log_{10}(d)$

– with $f = 800\text{MHz}, L_{dB}(d) = 127.3 + 35.22 \times \log_{10}(d)$

– with $f = 700\text{GHz}, L_{dB}(d) = 125.3 + 35.22 \times \log_{10}(d)$

Using these formulas, the maximum range of the cell will be, when there is no traffic in the cell:

| Characteristic | Unit | f = 800 MHz | f = 2.6 GHz |
|---|---|---|---|
| $PL_{QPSK}$ | dB | 144.5 | 144.5 |
| $d_{max}$ | M | 3,560 | 990 |
| $PL_{16QAM}$ | dB | 138 | 138 |
| $d_{max}$ | M | 2,330 | 650 |
| $PL_{64QAM}$ | dB | 131 | 131 |
| $d_{max}$ | M | 1,470 | 410 |

If the cell is serving 25 active mobile terminals and $f = 2.6$ *GHz*, one RB uplink and one RB downlink per mobile, the result is as follows. Note that the limiting direction is the downlink.

| Characteristic | Unit | Uplink | Downlink | dmax (m) |
|---|---|---|---|---|
| EIRP | dBm<br>W | 21<br>0.125 | 33<br>2 | |
| Rec. Thres. $_{QPSK}$ | dBm<br>W | −126.5<br>$2.2\ 10^{-16}$ | −106.5<br>$2.2\ 10^{-14}$ | 585 m |
| PL $_{QPSK}$ | dB | 144.5 | 139.5 ⇒136.5 | 585 m |
| Rec. Thres. $_{16QAM}$ | dBm<br>W | −120<br>$10^{-15}$ | −100<br>$10^{-13}$ | 380 m |
| PL $_{16QAM}$ | dB | 138 | 133 ⇒ 130 | 380 m |
| Rec. Thres. $_{64QAM}$ | dBm<br>W | −113<br>$5\ 10^{-15}$ | −93<br>$5\ 10^{-13}$ | 240 m |
| PL $_{64QAM}$ | dB | 131 | 126 ⇒ 123 | 240 m |

Taking into account the principal frequency bands to carry LTE, with the same 25 active UE, granted one RB uplink and one RB downlink, the results are as follows:

| Characteristic | unit | f = 700 MHz | f = 800 MHz | f = 1.8 GHz | f = 2.6 GHz |
|---|---|---|---|---|---|
| PL $_{QPSK}$ | dB | 144.5 | 144.5 | 144.5 | 144.5 |
| $d_{max}$ | m | 3,500 | 3,100 | 1,400 | 990 |
| PL $_{16QAM}$ | dB | 138 | 138 | 138 | 138 |
| $d_{max}$ | m | 2,300 | 2,000 | 920 | 650 |
| PL $_{64QAM}$ | dB | 131 | 131 | 131 | 131 |
| $d_{max}$ | m | 1,450 | 1,300 | 585 | 410 |

Now, let us take five active mobile terminals and $f = 2.6\ GHz$, five RB uplink and 25 RB downlink per mobile. The limiting factor is the downlink. Note the very low radiated power for the UE.

| Characteristic | Unit | Uplink | Downlink | dmax (m) |
|---|---|---|---|---|
| EIRP | dBm<br>W | 14<br>0.025 | 33<br>2.5 | |
| Rec. Thres. $_{QPSK}$ | dBm<br>W | −126.5<br>$2.2\ 10^{-16}$ | −106.5<br>$2.2\ 10^{-14}$ | 585 m |
| PL $_{QPSK}$ | dB | 140.5 ⇒ 137.5 | 136.5 | 585 m |
| Rec. Thres. $_{16QAM}$ | dBm<br>W | −120<br>$10^{-15}$ | −100<br>$10^{-13}$ | 380 m |
| PL $_{16QAM}$ | dB | 134 ⇒ 131 | 130 | 380 m |
| Rec. Thres. $_{64QAM}$ | dBm<br>W | −113<br>$5\ 10^{-15}$ | −93<br>$5\ 10^{-13}$ | 240 |
| PL $_{64QAM}$ | dB | 127 ⇒ 124 | 123 | 240 |

And again, the results obtained in the four frequency bands with the same configuration of UE (five active mobiles granted each five RB uplink and 25 RB downlink).

| Characteristic | unit | f = 700 MHz | f = 800 MHz | f = 1.8 GHz | f = 2.6 GHz |
|---|---|---|---|---|---|
| $PL_{QPSK}$ | dB | 136.5 | 136.5 | 136.5 | 136.5 |
| $d_{max}$ | m | 2,100 | 1,820 | 840 | 585 |
| $PL_{16QAM}$ | dB | 130 | 130 | 130 | 130 |
| $d_{max}$ | m | 1,360 | 1,200 | 550 | 380 |
| $PL_{64QAM}$ | dB | 123 | 123 | 123 | 123 |
| $d_{max}$ | m | 860 | 750 | 350 | 240 |

## 1.12. Appendix 2: relaying the radio links

Radio relays have been used to enhance the coverage in areas, which are out of the range of the base stations.

The direct link between the eNodeB and the UE can be modeled with a simple formula for rough calculations:

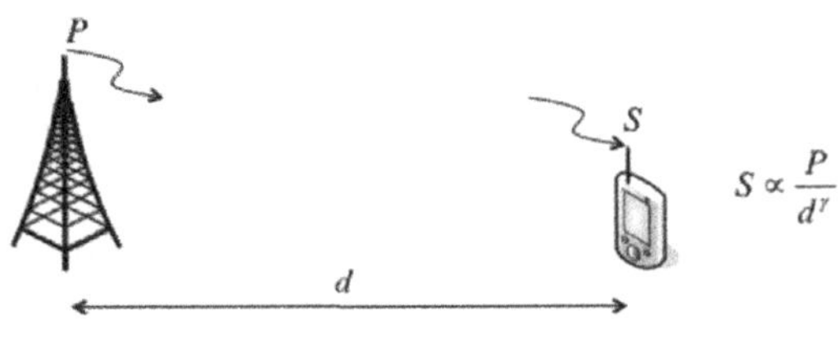


With the introduction of a relay, the schema becomes, again with a simplified modeling:

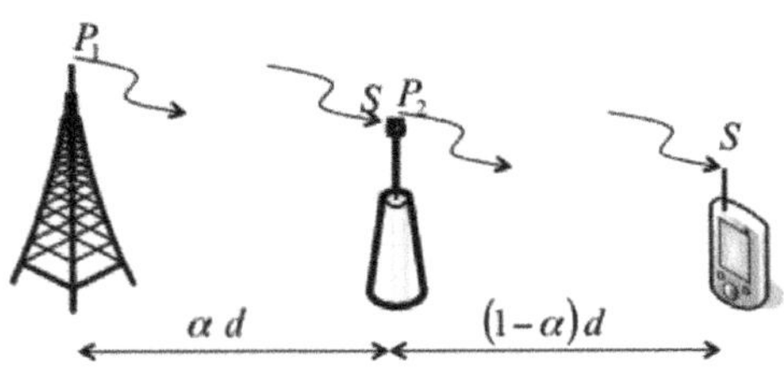

$$S\alpha \frac{P_1}{(\alpha d)^{\gamma}} = \frac{P_2}{((1-\alpha)d)^{\gamma}}$$

LTE standard introduces relaying in order to diminish the radiated power, with:

$$P_1 + P_2 < P$$

with the simplified propagation model:

$$\Rightarrow \alpha^{\gamma} + (1-\alpha)^{\gamma} < 1$$

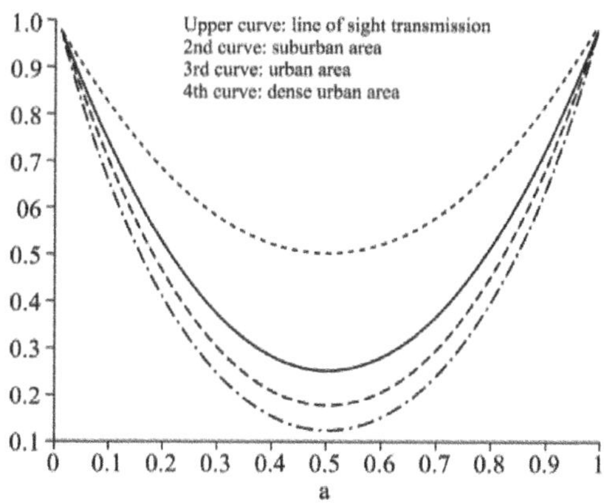


*Upper curve: line of sight transmission;*

*2nd curve: suburban area;*

*3rd curve: urban area;*

*4th curve: dense urban area.*

When the UE and the relay have the same sensitivity, the optimal position of the relay is in the middle of the distance between eNodeB and UE.

Relaying makes more sense in a dense urban environment.

Issue: if the relay is half duplex, the network capacity is divided by two.

Example of calculation of the impact of inserting a relay.

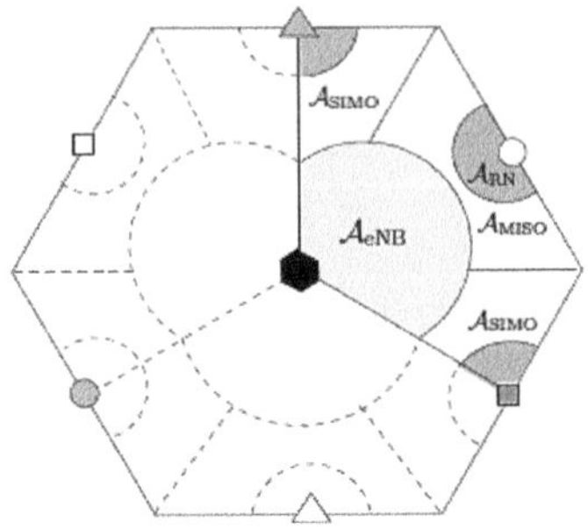


where:

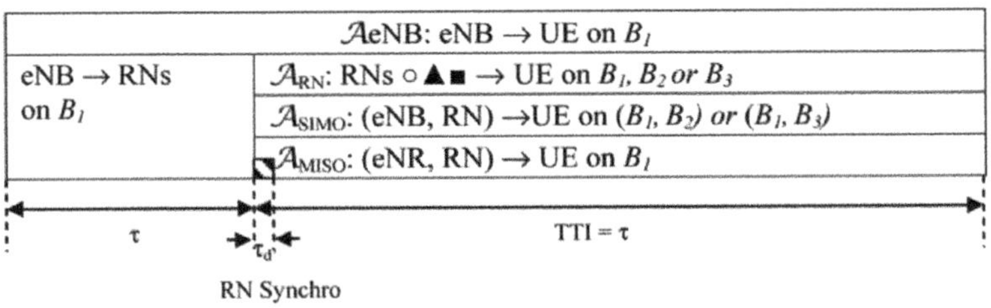


| Parameter | Value |
|---|---|
| Carrier frequency<br>Frequency reuse pattern<br>Channel bandwidth<br>Subcarrier bandwidth | 2.6 GHz<br>$1 \times 3 \times 1$<br>10 MHz<br>B = 15 kHz |
| eNB - RN propagation (LOS)<br>eNB- UE paopagation (NLOS)<br>RN-UE propagation (NLOS)<br>Shadowing deviation | $\alpha = 10^{-10.2}$ . $\gamma = 2.2$<br>$\alpha = 10^{-13.9}$ . $\gamma = 3.9$<br>$\alpha = 10^{-14.3}$ . $\gamma = 3.95$<br>$\sigma_x = 6$ dB |
| eBN height<br>eNB site-to-site distance<br>eNB-UE minimal distance<br>eNB max Tx power per sector<br>eNB antenna pattern (dBi) | 25 m (above rooftop)<br>$d = 2\sqrt{3}$km<br>$d_{min} = 30$ m<br>46 dBm<br>$G_{eNB}(\theta^*) = 17 - \min\left(12\frac{\theta^2}{70^2}, 20\right)$ |
| RN height<br>RN-UE minimal distance<br>RN max Tx power<br>RN antenna pattern (dBi) | 22 m (above rooftop)<br>$d_{min} = 30$ m<br>36 dBm<br>$G_{RN} = 7$ (omnidirectionnal) |

| | No relay | Multi region |
|---|---|---|
| Cell throughput $C$ (Mbps) | 24.75 | 24.96 |
| eNB reg throughput $\bar{C}_{eNB}$ (Mbps) | – | 8.53 |
| Energy (Joules) | 0.0398 | 0.0394 |

In the eNodeB region, the spectral efficiency is as follows:

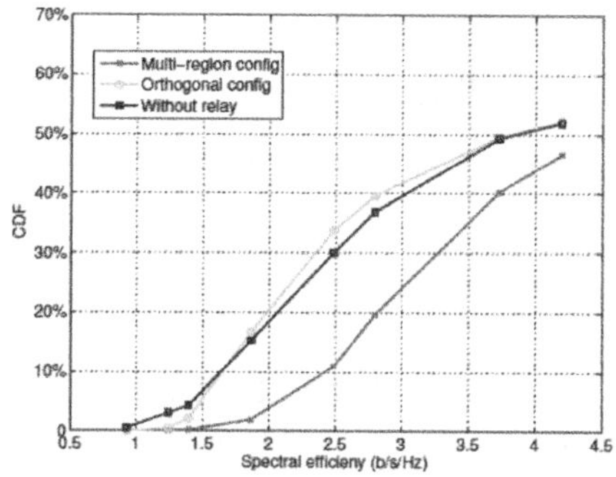


In the relay region it becomes:

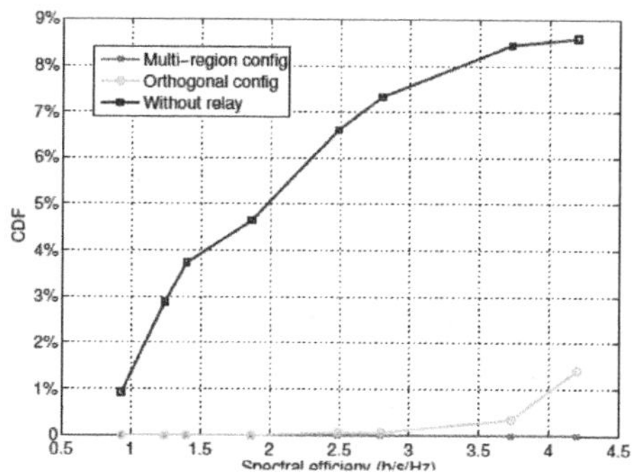


And in the diversity region:

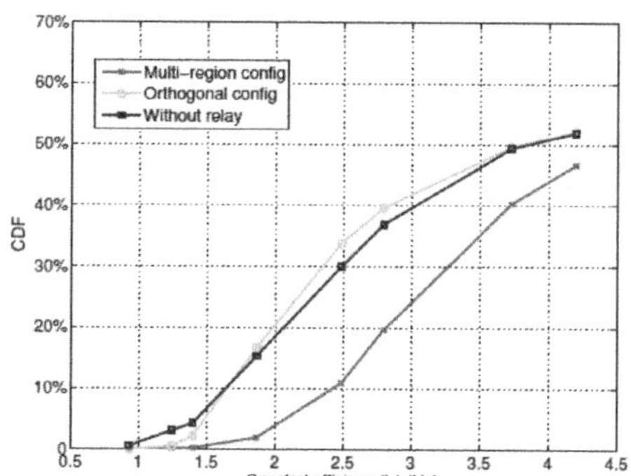

The relay:

– increases spectral efficiency;

– increases average cell capacity;

– diminishes transmit energy dissipation.

## 1.13. Appendix 3: LTE-Advanced: requirements

3GPP decided to further enhance LTE not only to qualify as a 4G technology but also to surpass it. In order to do so, it defined the following requirements:

– increased peak data rates (Gbps):

| Low mobility scenario | High mobility scenario |
|---|---|
| 1 | 0.1 |

| Downlink | Uplink |
|---|---|
| 1 | 0.5 |

– improved cell edge throughput:

| Direction | Antenna configuration | Bps/Hz/user |
|---|---|---|
| UL | 1 × 2 | 0.04 |
| UL | 2 × 4 | 0.07 |
| DL | 2 × 2 | 0.07 |
| DL | 4 × 2 | 0.09 |
| DL | 4 × 4 | A.12 |

– improved spectrum efficiency: peak (bps/Hz):

| Antenna configuration | Spectrum efficiency | Downlink | Uplink |
|---|---|---|---|
| | Peak | 1 bps/Hz | 0.5 bps/Hz |
| UL 1×2 | Average | | 1.2 bps/Hz/cell |
| UL 2×4 | Average | | 2.0 bps/Hz/cell |
| DL 2 × 2 | Average | | 2.4 bps/Hz/cell |
| DL 4 × 2 | Average | | 2.6 bps/Hz/cell |
| DL 4 × 4 | Average | | 3.7 bps/Hz/cell |

– spectrum flexibility: new spectrum bands are available (in addition release 8):

| Under 1 GHz | Over 1 GHz |
|---|---|
| 450–470 MHz | 2.3–2.4 GHz |
| 698–862 MHz | 3.4–4.2 GHz |
| 790–862 MHz | 4.4–4.99 GHz |

In addition to the bandwidths of release 8, LTE-A should support wider bandwidth allocations of up to 100 MHz, possible aggregating contiguous and/or non-contiguous spectrum. Also, it should support both unpaired (TDD) and paired (FDD) spectrum allocations:

– interworking: should provide better or at least the same performance of Release 8;

– mobility:

| Speed (km/h) | Support |
|---|---|
| 0–10 | Enhanced |
| 10–350 | Preferably enhanced |

In addition to the previous requirements, LTE-A is targeted to have low cost and complexity UE and infrastructure, enhanced support for MBMS and VoIP, and consider deployment scenarios for indoor eNodeBs.

# 2

# Operation and Maintenance

## 2.1. Introduction

Operation and maintenance (O&M) is one of the most important tasks of the network operator. Initially, manufacturers put a lot of energy to develop their own O&M tools. That strategy aimed at trapping the operators and preventing them from changing to more efficient products. But in the end, this behavior did not prevent the collapse of quite a few actors.

On its side, from the beginning of its activity, 3rd Generation Partnership Project (3GPP) took the challenging decision to standardize telecommunication management. This is the task of the working group SA5 of TSG system aspects (SA).

SA working group (WG5) will specify the requirements, architecture and solutions for provisioning and management of the network (Radio Access Network (RAN), Core Network (CN), IP Multimedia Subsystem (IMS)) and its services. The WG will define charging solutions in alignment with the related charging requirements developed by the relevant WGs, and will specify the architecture and protocols for charging of the network and its services.

The WG will ensure its work is also applicable to the management and charging of converged networks, and potentially applicable to fixed networks. The WG will coordinate with other 3GPP WGs and all relevant standards developing organizations (SDOs) to achieve the specification work relevant to the provisioning, charging and management of the network and its services.

SA WG5 works on a huge number of TS. Anyone especially interested in this subject can find the list of TS in the appendix to this chapter (section 2.4). It seems difficult to summarize this work in a few pages. Interesting documents are in the 28 series, with standards for self-organizing networks, for Evolved UMTS Terrestrial Radio Access Network (E-UTRAN) management or Home Node B surveillance. The 32 series focuses on fault management, etc.

The principle of O&M is simple – it provides simple tools on work stations:

– which will deliver information on the behavior of each network element of the mobile network, as well as the general performance of subsystems;

– which will provide tools for taking efficient action against troubles in the network, eventually offering distant troubleshooting, either by switching from a faulty element to a spare one or by taking some action to repair it, e.g. by downloading through a data line a corrected piece of software or corrected parameters.

For O&M, the operator will invest duplicated control rooms where tens of technicians and engineers will monitor the network 24 hours a day, 7 days a week. They will take action to quickly solve dysfunctions appearing here and there in distant locations.

O&M will be efficient if the information flow follows standards, especially for displaying the items. Therefore, SA5 favors eXtensible Markup Language (XML) or Simple Object Access Protocol (SOAP) as the basis of the tools.

When the installed network does not (yet) offer the kind of tools that are necessary to master the situations, the operator can use two processes that will enlighten the effective situation:

– load tests;

– monitoring of standardized interfaces.

## 2.2. Load tests

### 2.2.1. *Dimensioning of network elements to smoothly carry the traffic*

Load tests may be necessary to verify that the network will not break down in certain circumstances. The dimensioning of the network elements has to be suitable for the load, which the network will have to carry. To avoid such difficulties, the Internet recommends the transmission of packets with Transmission Control Protocol (TCP)/Internet Protocol (IP) protocol. As soon as the packets are lost, TCP/IP drastically reduces the rhythm of packet sending before beginning to increase back the transmitted bitrate up to the situation. Some services, nevertheless, are transmitting with UDP, which does not have such a security.

Unlike telephony, where the statistical modeling of calls was set by Erlang more than 100 years ago, the Internet traffic is modeled along many different schemes, which are each given the name of an animal (e.g. elephant, mouse, etc.). These models can be implemented on computers and can provide a means to calculate the size of the telecommunication ducts for the Internet. What seems to come out of experience in the field is that the individual connection may show significant peaks of traffic, especially on the downlink when the subscriber is downloading heavy files; so for customer satisfaction, it is recommended that the network dimensioning offers a high bitrate connection. However, experts consider that the dimensioning of the backbone network is not as colossal as the addition of all individual peak traffics. With millions of subscribers, there is a kind of statistical smoothing of the load, and the network will not have to lay down enormous transmission capacities. The model comes back to some simple Gaussian calculation with a mean value, which is surprisingly low (numbers are proprietary), and a standard deviation, which is the important information to be mastered since it may be quite important, especially taking into account differences between low traffic at some days in the year and peaks happening at some others.

In order to cope with discrepancies in the offered traffic, manufacturers have the possibility of installing large memories, which will help in staggering the peak of traffic and fluidize the flow of packets by delaying some of them while others fill the capacity of the telecommunication resource. The dimensioning of such buffers has to be consistent with the size of the observed peaks. In order to evaluate the appropriate memory to be included, the manufacturer has to know not only the number of gigabytes but also the number of read/write operations to be provided. Some cheap components offer adequate capacity for the memory, but cannot stand up to thousands or millions of read/write actions.

Fixed operators, which hold an important customer base for ADSL or other fast Internet services are in a position to install analyzers on their lines, which will record the offered traffic for, say, one month at a time of the year when the traffic is known to be heavy. Thus, they can build a good knowledge of the traffic shape as it is in the real life. Of course, Long Term Evolution (LTE) traffic may differ more or less, but since most of the applications are both on ADSL (or equivalent) and on mobile Internet, the approach will not be too much erroneous. For them, it is a means to validate the conception of the network equipment they receive from manufacturers: they can easily reproduce the observed traffic with traffic generators and inject it into the piece of hardware/software proposed for validation. Such load tests have to be carried out for a long period, e.g. a few days. This precaution will avoid us suffering later in effective field operation from buffer disorders due to inappropriate components.

### 2.2.2. *Dimensioning of signaling channels*

LTE, like all modern mobile systems, manages the traffic with a lot of signaling, both on the radio path and in the core network. In the core network, resources are available to increase the capacity of signaling ducts, but the manufacturer's design may have limited the maximum limit to under what could be necessary to have the flow of communications to and from the mobiles. On the radio path, the dimensioning is even more delicate, because the resource is shared with the communications' content.

For signaling, a good model can be set using the Aloha and Slotted-Aloha processes. Let us assume that the message to be carried by one signaling channel has a duration of *P*' seconds and appears at instant *t*. This message will be correctly delivered if no other message is coming before ($t + P$'). If there is a collision, the transmission of the message must be reiterated.

Not knowing the rate at which messages are incoming to this signaling channel, we consider that they follow Poisson's law, including the reiterations, which can be spread randomly by the system – which increases the value of choosing Poisson's law. In real life and in order to reach accurate simulation, a protocol analyzer has to be set on the channel under study, which will record the effective arrival of signaling messages and their reiterations.

λ being the rate of Poisson's process, the probability that *k* messages arise during the *P*' timeslot is:

$$Pk(t) = \frac{(\lambda p')^k}{k!} e^{-\lambda p'}$$

Poisson's process has no memory; the only parameter is the duration of the timeslot on which events are observed. If there is no limit to the *k* number of messages, the probability of occurrences is:

$$Pr = e^{-\lambda p'} . e^{\lambda p'} = e^{-2\lambda p'}$$

If *k* is limited by some mechanism in the telecommunication system:

$$PR(k) = 1 - (Pc)^k$$

where *Pc* is the probability of a collision, which gives rise to reiteration.

$$PR(k) = 1 - (1 - e^{2\lambda p'})^k$$

The mean value of *k* is:

$$< k >= e^{-2\lambda P'}(1-(1-e^{-2\lambda P'})^{k}))$$

The optimum is reached for

$$e^{-\lambda P'} + (-2\lambda P' e^{\uparrow}(-2\lambda P')) = 0$$

$$\lambda P' = \frac{1}{2}$$

For an Aloha channel, the maximum throughput is limited to 0.5 Erlang. If $k \to \infty, < k > \to e$:

for *k* limited to 3, *PR*(3) = 0.747;

for *k* limited to 5, *PR*(5) = 0.899.

To increase efficiency, Aloha must be abandoned to the advantage of Slotted-Aloha. In the Slotted-Aloha process, messages are transmitted in precise windows of length *P* with a synchronization mechanism making it so that two messages either completely collide on their whole length or do not collide at all. The length of the messages is identical and just a little shorter than the window *P*. The difference is necessary for taking care of the acquisition of the synchronization and transmitter power up and down.

With Slotted-Aloha, the probability of successful transmission is the probability of the event:

(no apparition of message on the timeslot $\subset 0, to \subset$) $and\_$(no other on $\subset 0, to\ P \subset$ )

to being any time inside the window.

Therefore, $\Pr = e^{-\lambda P}$ with the same calculations as for Aloha above.

The optimum is reached for $\lambda P = 1$ and $Pr = \frac{1}{e}$.

Let us introduce $\lambda'' = \frac{1}{Pe}$

$$PR(k) = 1 - (1 - \frac{1}{e})^k$$

$$< k >= e(1 - 1 - \frac{1}{e})^k)$$

The Slotted-Aloha process allows us to transmit a traffic of 1 Erlang, of course including reiterations and slightly diminished by the small difference between the size of the window and the effective size of the message packet. Slotted-Aloha shows the same behavior as Aloha except that all traffic parameters are doubled.

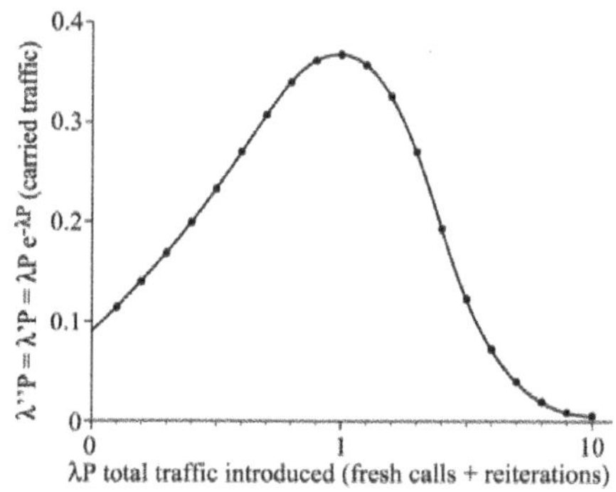


**Figure 2.1.** *Slotted-Aloha behavior*

This curve shows that the maximum fresh traffic (without reiterations) that can be transmitted is $\frac{1}{e}$ and is close to null when 10 reiterations are allowed.

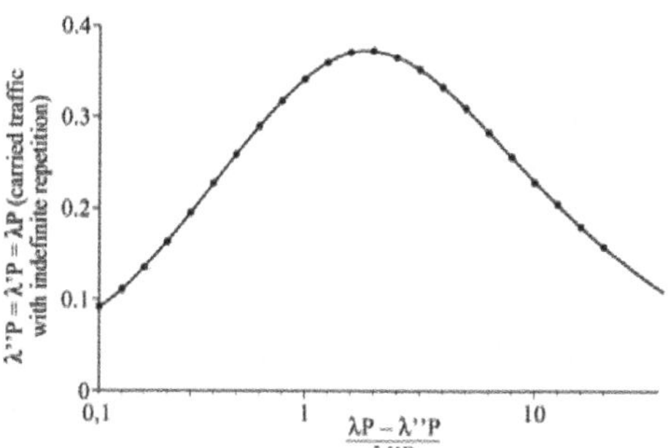


**Figure 2.2.** *Increase of repetitions as traffic increases*

The second curve shows the increasing size of the necessary buffer with a Slotted-Aloha process.

Looking at what happens in a stationary situation, the impact of repeated attempts to transmit one signaling packet is shown by the following curves.

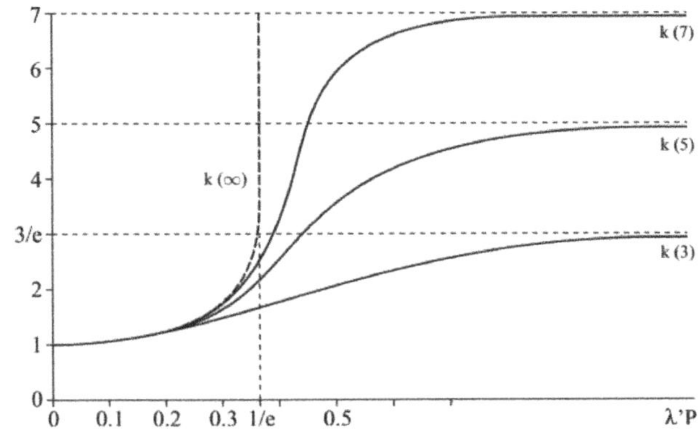


*Stationary situation*

For $k = \infty$, $PR = 1$ up to $\lambda' P = `a.3679$, but equals zero above this value.

When $k$ is limited, this abrupt behavior smooths down. The smaller the $k$, the smoother the answer.

For $k = 1$ (no repeat), $\lambda' = \lambda$ and $PR(1) = 1 - 6^{-\lambda P}$.

When the network is overcharged, protocols with a high number of reiterations collapse as soon as the influx of fresh messages rises above 0.5 Erlang. Values of 3 to 5 new trials seem a good compromise.

Of course, in case of discarding messages when the maximum number of attempts is reached, the throughput has to take into account lost messages.

The repetition of attempts is generally driven by interleaving them with a temporization taking random durations, which are chosen among a set of values. According to Kleinrock, the number of these values must exceed 16 for a correct operation.

If we assume that the number of devices attempting to be served is high, which is normally the case in an LTE cell, again Poisson's process may apply, which simplifies calculations.

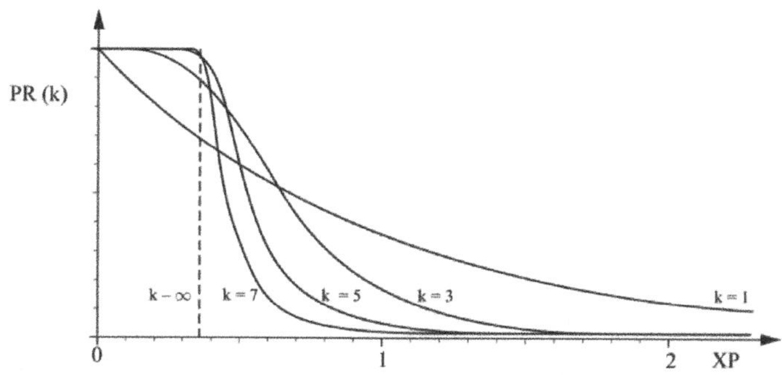


Considering *Ns* = *S* and *Np* = *Q* the intensity of Poisson's process, respectively, when there is no repetition (beginning of the flow of attempts) and when repetitions are increasing, the combined intensity when *r* terminals are waiting is:

$$X = (1-r)S + rQ$$

The proportion of waiting terminals *r* varies with the time when repetitions occur. *Na*(*r*) being the variation of the proportion of waiting terminals.

$$[(1-\exp(-X))]^k$$

The first term is the Slotted-Aloha term and the second term relates to lost messages.

The system is stable when Na (r) = 0 and $(d(Na(r))/dr < 0$.

For $k = 1$, $Na(r)$ is negative, so the system is stable. It is also the case after some time for finite values of $k$.

When $k$ is infinite (true Slotted-Aloha), the process diverges as soon as $r > 0.027$.

For $k$ taking the values of 3, 5 and 7, the equilibrium stands around $r = 0.01$.

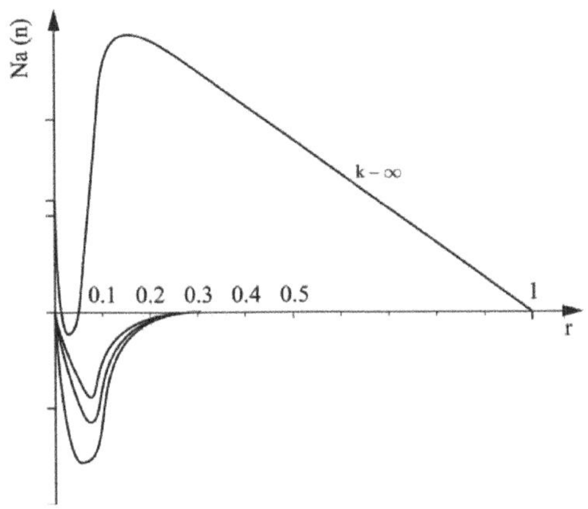


**Figure 2.3.** *r around 0, 1*

The following curves shed light on the behavior for low values of $r$.

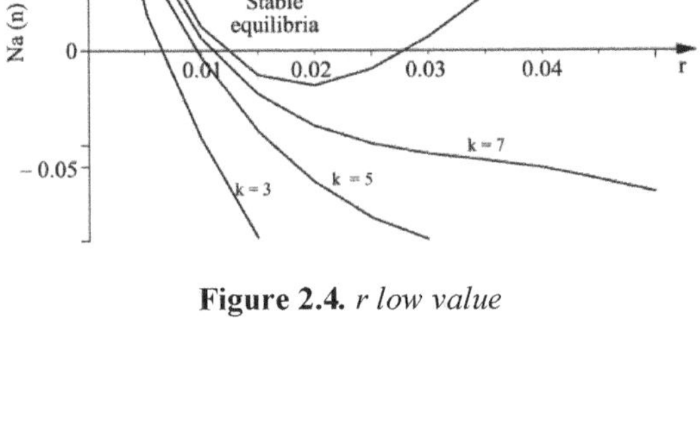


**Figure 2.4.** *r low value*

### 2.2.3. *Load tests on signaling channels*

The main method to check the resilience of a telecommunications network is to analyze the exchange of messages on the standardized interfaces. For these purposes, a few manufacturers have developed dedicated protocol analyzers, which own the full list of messages to be carried on these interfaces, which they are able to decode thanks to a very fast processing subsystem.

The load is generated from plenty of mobile terminals, which are manipulated in such a way that they generate a very significant flow of signaling messages on the chosen interface (e.g. S1, X2). Also, some message generators are made available, which make it much easier to manage.

The protocol analyzer shows the modification of the exchange of messages on the standardized interface and by these means, the behavior of the network design can eventually be controlled down to crash.

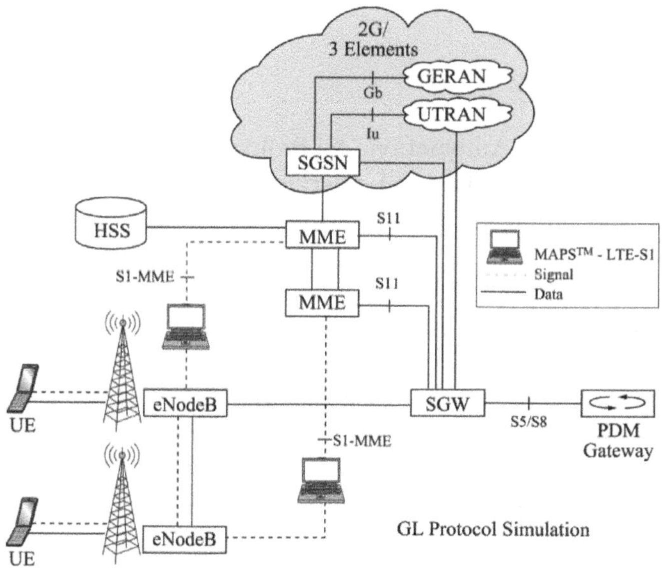


**Figure 2.5.** *Interfaces to be monitored*

## 2.3. Use of protocol analyzer: example of MAPS system

### 2.3.1. *Background*

The above diagram depicts some of the internals of the LTE network. The user equipment (UE) has a very high-speed radiofrequency (RF) connection to the nearest local tower. This high-speed connection is always on, as long as the mobile is powered up. LTE is an all IP infrastructure with built-in service priority – audio and video are given priority. All necessities such as IP address, authentication and security are validated. Instant resources over RF (the air) and IP (internal network) are made available depending on what the user is attempting to do. Also, LTE is designed for compatibility with older second generation (2G) and third generation (3G) mobile systems.

eNodeB is the base station in the LTE/System Architecture Evolution (SAE) S1 interface and also includes mobility management entities (MME) (to handle signaling of control plane) as shown in the network diagram above.

### 2.3.2. *Overview*

GL's Message Automation & Protocol Simulation (MAPS™) designed for testing LTE – S1 interface (any interface excluding air interface) can simulate eNodeB (Evolved Node B) and MME.

The application is available as:

– MAPS™ – LTE-S1 (Item # PKS140) for S1-MME interface.

MAPS™ is also capable of simulating S11 and S5/S8 (LTE-eGTP) interfaces. Refer to MAPS™ LTE eGTP (S3, S4, S5, S8, S10, S11 and S16) webpage for more details [GL 14]. Other interfaces such as S4, S11 and S12 will also be supported by the application in the near future.

The application gives users the unlimited ability to edit S1 Application Protocol (S1-AP)/Non-Access-Stratum (NAS) messages and call scenarios (message sequences). "Message sequences" are generated through scripts. “Messages” are created using message templates.

Possible applications include:

– simulating up to 500 smartphones (UEs) powering up and down;

– authenticating and confirming security procedures;

– quality of service (QoS) requests for greater or lesser bandwidth;

– temporarily addressing management for mobility and security.

The product also supports mobile traffic core – GTP (ETH101) simulation for user-plane packet transmission and reception services between any two nodes in LTE and UMTS networks. The mobile traffic core – gateway (ETH102) module allows simulation of gateway traffic to test media gateway telephony interfaces over IP. These modules also support generation and verification of data traffic such as email, File Transfer Protocol (FTP), web (Hyper Text Transfer Protocol (HTTP), video and more.

GL also provides a GUI-based LTE analyzer for online capture and decoding of the signaling in real time both during tests and as a stand-alone tracer for live systems.

### 2.3.3. *Main features*

– Simulates eNodeB, and MME elements in S1 interface.

– Supports LTE control plane.

– Generates and responds to hundreds of UE signaling (load testing).

– Generates and process S1/NAS messages (valid and invalid).

– Impairments can be applied to messages to simulate error conditions.

– Supports customization of call flows and message templates using script editor and message editor.

– Ready-to-use scripts for quick testing.

– Supports scripted call generation and automated call reception.

– Supports mobile traffic simulation with additional licenses.

– Provides protocol trace with full message decoding, and graphical ladder diagrams of call flow with time stamp.

– Script-based and protocol-independent software architecture.

– Provides call statistics and associated captured events and error events during call simulation.

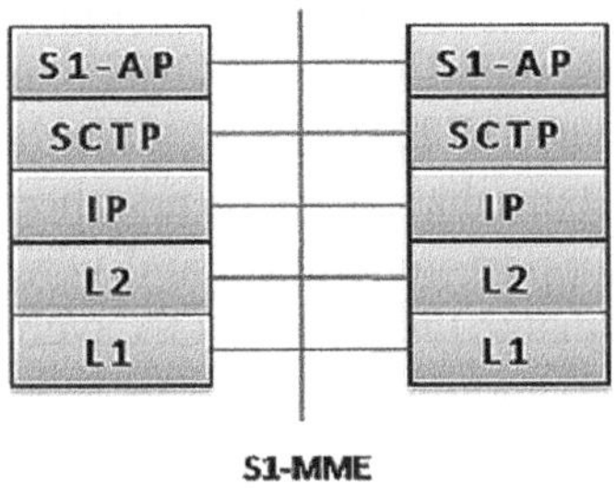


**Figure 2.6.** *Control plane for S1-MME interface*

### 2.3.4. *Supported protocol standards*

– *S1 Application Protocol (S1-AP):* application layer protocol between the eNodeB and the MME.

*Stream Control Transmission Protocol (SCTP) for the control plane (SCTP):* this protocol guarantees delivery of signaling messages between MME and eNodeB (S1). SCTP is defined in RFC 2960 [RFC 00].

| Protocol | Standard specification |
|---|---|
| *S1 Interface* | |
| S1 Application Protocol (S1-AP) | 3GPP 36.413 9.0.0 (2009-09) |
| Non-Access-Stratum (NAS) | 3GPP TS 24.301 V9.0.0 (2009-09) |

**Table 2.1.** *Standards*

## 2.3.5. *Test configuration*

A test bed setup is configured to use a predefined XML profile file. The multiple profiles created in the default profile (XML) selected in the test bed setup are accessible from the call generation and reception windows.

This feature allows users to configure the necessary parameters to establish communication between MAPS™ LTE S1 and the DUT. The S1 interface supports SCTP configuration.

For IP-based protocols, the MAPS™ test bed should be setup with mandatory source/destination port, address, stream ID and payload ID parameters to establish connection with DUT. These parameters enable the users to configure the MAPS™ application as the required entity in SCTP/ Rich Internet Application (IUA)/UDP/ Media Transfer Protocol. It is a Media Transfer Protocol (MTP) layers of IP network. The default profile set in the test bed setup is used to configure MAPS™ LTE S1 for call configuration.

*Typical call flow scenario*

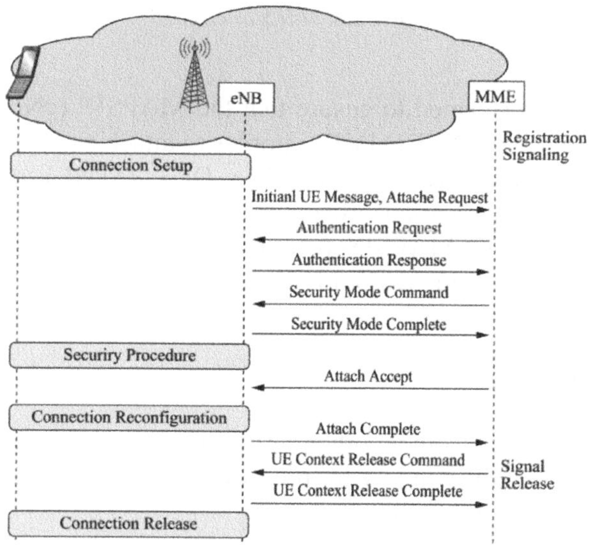


**Figure 2.7.** *LTE-S1 signaling scenario (messages between eNodeB (eNB) and MME)*

### 2.3.6. *Call generation*

The call generation option allows the user to simulate outgoing communications where an outgoing call is initiated by sending call control messages using proper scripts and profiles. The profile allows necessary parameters of call control messages to be changed during runtime.

#### 2.3.6.1. *Testing MME in S1 interface*

Here, MAPS™ acts as eNodeB and tests the DUT that is configured as MME. MAPS™ initiates the messages toward MME and responds to messages from MME (DUT).

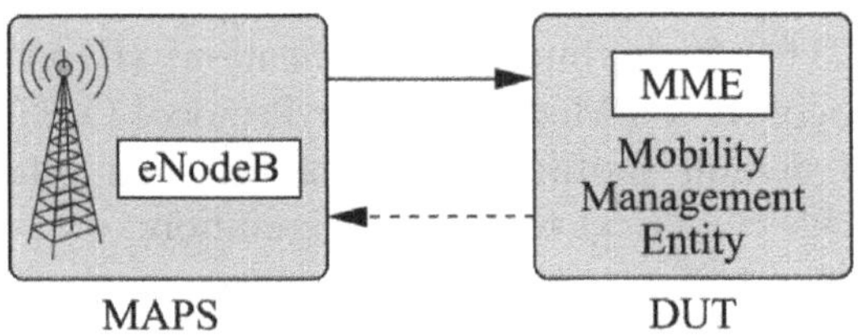


**Figure 2.8.** *MAPS as eNodeB*

The test is performed to ensure that the MAPS™ (eNodeB) sends an attach request to DUT (MME) in order to attach an Evolved Packet Core (EPC) for packet services. The DUT (MME) responds by initiating an authentication procedure and sending an authentication request toward UE. MAPS™ (eNodeB) receives the authentication request message as depicted in Figure 2.8.

### 2.3.7. *Call reception*

A call generated from an other entity can be automatically detected in the call reception window by presetting the required scripts in the incoming call handler window.

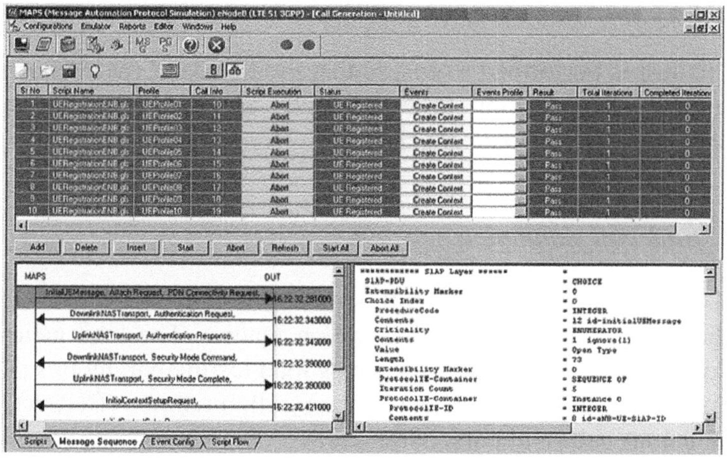


**Figure 2.9.** *Call generation at eNodeB node*

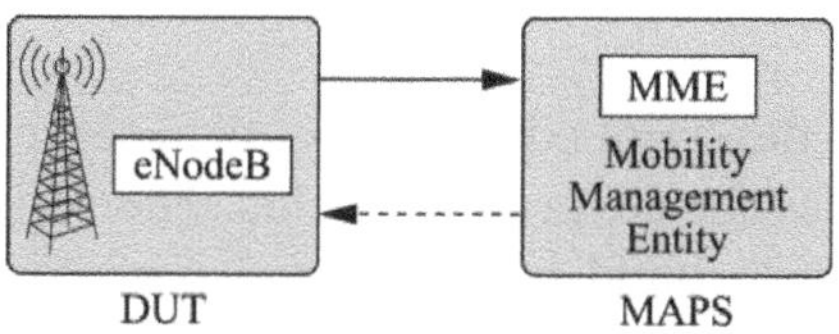


**Figure 2.10.** *MAPS as MME*

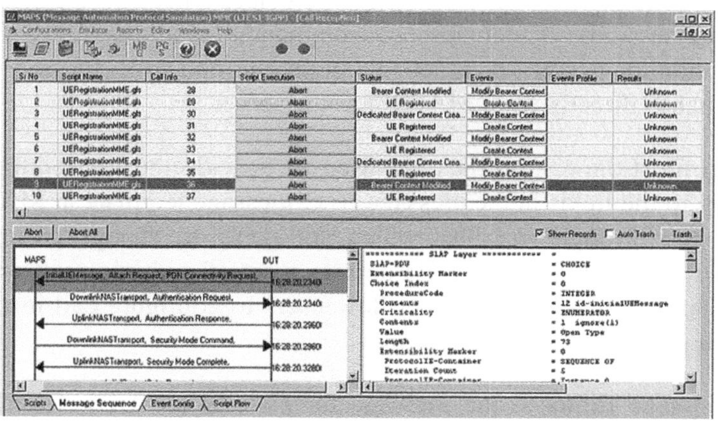


**Figure 2.11.** *Call reception at MME*

#### 2.3.7.1. *Testing eNodeB in S1 interface*

This test is performed to ensure that the MAPS™ (MME), in order to initiate authentication process, sends an authentication request message as response to an attach request from eNodeB (DUT).

### 2.3.8. *Bulk call simulation*

MAPS™ supports bulk call simulation and stress/load testing capabilities through the load generation feature. The load generation window helps users to configure stress/load testing parameters such as call per second (CPS) or busy hour call attempts (BHCA), max simultaneous calls and burst parameters.

### 2.3.9. *Customization of call flow and messages using preprocessing tools*

Script Editor – the script editor allows the user to create/edit scripts and define variables for the protocol fields. The script uses predefined message templates to build call flow and perform send and receive actions. The script editor provides options to run the test for multiple iterations in sequential or random flow. Commands allow retransmission of messages with specific intervals. It also includes traffic commands to send and monitor voice, tones and digits on the created sessions. It includes raw commands (send/monitor signaling bits, monitor power level, set idle code and end task) to send WCS commands directly from MAPS™ to the server.

Profile Editor – profiles are used to provide the user-configured values to the fields in the messages (i.e. message template in MAPS™ LTE S1) through variables which are going to change during the course of a call. An XML file defines a set of multiple profiles with varying parameter values that allow users to configure call instances in call generation and receive calls.

Message Editor – the message template is a *.HDL file that is composed of protocol-encoding parameters with preset values. It is required to create a message template for every message in a protocol.

The message templates are called within the scripts to perform scenario-based testing.

### 2.3.10. *Call flow and script execution control*

Message sequence – MAPS™ LTE S1 provides protocol trace with full message decoding, custom trace and graphical ladder diagrams of call flow with time stamp while simulation is running. The call flow graph allows easy verification of the messages exchanged between MAPS™ LTE S1 and DUT. See call generation and call reception for details.

Script contents and script flow – the scripts contents window displays the contents of the script selected for call generation or reception. The script flow window displays the set of statements successfully executed by MAPS™ – LTE S1 to help users in troubleshooting a particular call scenario.

Events and event profile editor – user-defined events allow redirection of script execution on-the-go. The custom parameters in the events can also be changed during script execution using Event profiles.

### 2.3.11. *Call statistics, events, link status*

Call statistics and status – by default, all call handling scripts (irrespective of the type of the functions) are assessed by MAPS™ LTE S1 to provide statistical information about total calls, active calls, completed calls, past calls and failed calls. It is also possible to characterize the statistical information under different groups of call handling scripts under a unique heading.

In addition, call generation and call reception windows provide useful call status and script execution results.

Events reporting – MAPS™ LTE S1 provides event log, error events and captured errors windows that log the captured events and errors encountered during the progress of the call.

## 2.4. Appendix: TS of SA5 working group of 3GPP TSG SA

| | |
|---|---|
| TS 12.00 | Objectives and Structure of GSM Public Land Mobile Network (PLMN) Management |
| TS 12.01 | Network Management (NM); Part 2: Common aspects of SM/DCS 1800 Network Management |
| TS 12.02 | Subscriber, Mobile Equipment (ME) and Services Data Administration |
| TS 12.03 | Security Management |
| TS 12.04 | Performance data measurements |
| TS 12.05 | Subscriber Related Call and Event Data |
| TS 12.06 | Network Configuration Management and Administration |
| TS 12.08 | Subscriber and Equipment trace |
| TS 12.10 | Maintenance Provisions for Operational Integrity of MSs |
| TS 12.11 | Fault management of the Base Station System (BSS) |
| TS 12.15 | General Packet Radio Service (GPRS); GPRS Charging |
| TS 12.20 | Base Station System (BSS) Management Information |
| TR 12.30 | ETSI Object Identifier Tree; Mobile Domain O&M |
| TS 12.71 | Location Services (LCS); Location services management |
| TS 28.390 | Telecommunication management; Fixed Mobile Convergence (FMC); Interface Integration Reference Point (IRP) Solution Profiles (SPs) |

| | |
|---|---|
| TS 28.401 | Telecommunication management; Performance Management (PM); Performance measurements for Core Network (CN) and non-3GPP access Interworking System |
| TS 28.402 | Telecommunication management; Performance Management (PM); Performance measurements for Evolved Packet Core (EPC) and non-3GPP access Interworking System |
| TS 28.403 | Performance Management; performance measurements for WLAN |
| TS 28.601 | Telecommunication management; Core Network (CN) and non-3GPP access interworking system Network Resource Model (NRM) Integration Reference Point (IRP); Requirements |
| TS 28.602 | Telecommunication management; Core Network (CN) and non-3GPP access interworking system Network Resource Model (NRM) Integration Reference Point (IRP); Information Service (IS) |
| TS 28.606 | Telecommunication management; Core Network (CN) and non-3GPP access interworking system Network Resource Model (NRM) Integration Reference Point (IRP); Solution Set (SS) definitions |
| TS 28.611 | Telecommunication management; Evolved Packet Core (EPC) and non-3GPP access interworking system Network Resource Model (NRM) Integration Reference Point (IRP); Requirements |
| TS 28.612 | Telecommunication management; Evolved Packet Core (EPC) and non-3GPP access interworking system Network Resource Model (NRM) Integration Reference Point (IRP); Information Service (IS) |
| TS 28.616 | Telecommunication management; Evolved Packet Core (EPC) and non-3GPP access interworking system Network Resource Model (NRM) Integration Reference Point (IRP); Solution Set (SS) definitions |
| TS 28.620 | Telecommunication management; Fixed Mobile Convergence (FMC) Federated Network Information Model (FNIM) Umbrella Information Model (UIM) |

| | |
|---|---|
| TS 28.621 | Telecommunication management; Generic Network Resource Model (NRM) Integration Reference Point (IRP); Requirements |
| TS 28.622 | Telecommunication management; Generic Network Resource Model (NRM) Integration Reference Point (IRP); Information Service (IS) |
| TS 28.623 | Telecommunication management; Generic Network Resource Model (NRM) Integration Reference Point (IRP); Solution Set (SS) definitions |
| TS 28.624 | Telecommunication management; State management data definition Integration Reference Point (IRP); Requirements |
| TS 28.625 | Telecommunication management; State management data definition Integration Reference Point (IRP); Information Service (IS) |
| TS 28.626 | Telecommunication management; State management data definition Integration Reference Point (IRP); Solution Set (SS) definitions |
| TS 28.627 | Telecommunication management; Self-Organizing Networks (SON) Policy Network Resource Model (NRM) Integration Reference Point (IRP); Requirements |
| TS 28.628 | Telecommunication management; Self-Organizing Networks (SON) Policy Network Resource Model (NRM) Integration Reference Point (IRP); Information Service (IS) |
| TS 28.629 | Telecommunication management; Self-Organizing Networks (SON) Policy Network Resource Model (NRM) Integration Reference Point (IRP); Solution Set (SS) definitions |
| TS 28.631 | Telecommunication management; Inventory Management (IM) Network Resource Model (NRM) Integration Reference Point (IRP); Requirements |
| TS 28.632 | Telecommunication management; Inventory Management (IM) Network Resource Model (NRM) Integration Reference Point (IRP); Information Service (IS) |

| | |
|---|---|
| TS 28.633 | Telecommunication management; Inventory Management (IM) Network Resource Model (NRM) Integration Reference Point (IRP); Solution Set (SS) definitions |
| TS 28.651 | Telecommunication management; Universal Terrestrial Radio Access Network (UTRAN) Network Resource Model (NRM) Integration Reference Point (IRP); Requirements |
| TS 28.652 | Telecommunication management; Universal Terrestrial Radio Access Network (UTRAN) Network Resource Model (NRM) Integration Reference Point (IRP); Information Service (IS) |
| TS 28.653 | Telecommunication management; Universal Terrestrial Radio Access Network (UTRAN) Network Resource Model (NRM) Integration Reference Point (IRP); Solution Set (SS) definitions |
| TS 28.654 | Telecommunication management; GSM/EDGE Radio Access Network (GERAN) Network Resource Model (NRM) Integration Reference Point (IRP); Requirements |
| TS 28.655 | Telecommunication management; GSM/EDGE Radio Access Network (GERAN) Network Resource Model (NRM) Integration Reference Point (IRP); Information Service (IS) |
| TS 28.656 | Telecommunication management; GSM/EDGE Radio Access Network (GERAN) Network Resource Model (NRM) Integration Reference Point (IRP); Solution Set (SS) definitions |
| TS 28.657 | Telecommunication management; Evolved Universal Terrestrial Radio Access Network (E-UTRAN) Network Resource Model (NRM) Integration Reference Point (IRP); Requirements |
| TS 28.658 | Telecommunication management; Evolved Universal Terrestrial Radio Access Network (E-UTRAN) Network Resource Model (NRM) Integration Reference Point (IRP); Information Service (IS) |
| TS 28.659 | Telecommunication management; Evolved Universal Terrestrial Radio Access Network (E-UTRAN) Network Resource Model (NRM) Integration Reference Point (IRP); Solution Set (SS) definitions |

| | |
|---|---|
| TS 28.661 | Telecommunication management; Generic Radio Access Network (RAN) Network Resource Model (NRM) Integration Reference Point (IRP); Requirements |
| TS 28.662 | Telecommunication management; Generic Radio Access Network (RAN) Network Resource Model (NRM) Integration Reference Point (IRP); Information Service (IS) |
| TS 28.663 | Telecommunication management; Generic Radio Access Network (RAN) Network Resource Model (NRM) Integration Reference Point (IRP); Solution Set (SS) definitions |
| TS 28.667 | Radio Planning Tool (RPT) interface NRM IRP – Requirements |
| TS 28.668 | Radio Planning Tool (RPT) interface NRM IRP – Information Service (IS) |
| TS 28.669 | Radio Planning Tool (RPT) interface NRM IRP – Solution Set (SS) definitions |
| TS 28.671 | Telecommunication management; Home Node B (HNB) Subsystem (HNS) Network Resource Model (NRM) Integration Reference Point (IRP); Requirements |
| TS 28.672 | Telecommunication management; Home Node B (HNB) Subsystem (HNS) Network Resource Model (NRM) Integration Reference Point (IRP); Information Service (IS) |
| TS 28.673 | Telecommunication management; Home Node B (HNB) Subsystem (HNS) Network Resource Model (NRM) Integration Reference Point (IRP); Solution Set (SS) definitions |
| TS 28.674 | Telecommunication management; Home enhanced Node B (HeNB) Subsystem (HeNS) Network Resource Model (NRM) Integration Reference Point (IRP); Requirements |
| TS 28.675 | Telecommunication management; Home enhanced Node B (HeNB) Subsystem (HeNS) Network Resource Model (NRM) Integration Reference Point (IRP); Information Service (IS) |

| | |
|---|---|
| TS 28.676 | Telecommunication management; Home enhanced Node B (HeNB) Subsystem (HeNS) Network Resource Model (NRM) Integration Reference Point (IRP); Solution Set (SS) definitions |
| TS 28.680 | WLAN Requirements |
| TS 28.682 | WLAN NRM; Information Service (IS) |
| TS 28.701 | Telecommunication management; Core Network (CN) Network Resource Model (NRM) Integration Reference Point (IRP); Requirements |
| TS 28.702 | Telecommunication management; Core Network (CN) Network Resource Model (NRM) Integration Reference Point (IRP); Information Service (IS) |
| TS 28.703 | Telecommunication management; Core Network (CN) Network Resource Model (NRM) Integration Reference Point (IRP); Solution Set (SS) definitions |
| TS 28.704 | Telecommunication management; IP Multimedia Subsystem (IMS) Network Resource Model (NRM) Integration Reference Point (IRP); Requirements |
| TS 28.705 | Telecommunication management; IP Multimedia Subsystem (IMS) Network Resource Model (NRM) Integration Reference Point (IRP); Information Service (IS) |
| TS 28.706 | Telecommunication management; IP Multimedia Subsystem (IMS) Network Resource Model (NRM) Integration Reference Point (IRP); Solution Set (SS) definitions |
| TS 28.707 | Telecommunication management; Evolved Packet Core (EPC) Network Resource Model (NRM) Integration Reference Point (IRP); Requirements |
| TS 28.708 | Telecommunication management; Evolved Packet Core (EPC) Network Resource Model (NRM) Integration Reference Point (IRP); Information Service (IS) |

| | |
|---|---|
| TS 28.709 | Telecommunication management; Evolved Packet Core (EPC) Network Resource Model (NRM) Integration Reference Point (IRP); Solution Set (SS) definitions |
| TS 28.731 | Telecommunication management; Transport Network (TN) interface Network Resource Model (NRM) Integration Reference Point (IRP); Requirements |
| TS 28.732 | Telecommunication management; Transport Network (TN) interface Network Resource Model (NRM) Integration Reference Point (IRP); Information Service (IS) |
| TS 28.733 | Telecommunication management; Transport Network (TN) interface Network Resource Model (NRM) Integration Reference Point (IRP); Solution Set (SS) definitions |
| TS 28.734 | Telecommunication management; Signalling Transport Network (STN) interface Network Resource Model (NRM) Integration Reference Point (IRP); Requirements |
| TS 28.735 | Telecommunication management; Signalling Transport Network (STN) interface Network Resource Model (NRM) Integration Reference Point (IRP); Information Service (IS) |
| TS 28.736 | Telecommunication management; Signalling Transport Network (STN) interface Network Resource Model (NRM) Integration Reference Point (IRP); Solution Set (SS) definitions |
| TS 28.751 | Telecommunication management; Subscription Management (SuM) Network Resource Model (NRM) Integration Reference Point (IRP); Requirements |
| TS 28.752 | Telecommunication management; Subscription Management (SuM) Network Resource Model (NRM) Integration Reference Point (IRP); Information Service (IS) |
| TS 28.753 | Telecommunication management; Subscription Management (SuM) Network Resource Model (NRM) Integration Reference Point (IRP); Solution Set (SS) definitions |

| | |
|---|---|
| TR 30.817 | Telecommunication management; Project scheduling and open issues for SA5, Release 7 |
| TR 30.818 | Telecommunication management; Project scheduling and open issues for SA5, Release 8 |
| TR 30.819 | Telecommunication management; Project scheduling and open issues for SA5, Release 9 |
| TR 30.820 | Telecommunication management; Project scheduling and open issues for SA5, Release 10 |
| TR 30.821 | Telecommunication management; Project scheduling and open issues for SA5, Release 11 |
| TS 32.005 | Telecommunications management; Charging management; 3G call and event data for the Circuit Switched (CS) domain |
| TS 32.015 | Telecommunications management; Charging management; 3G call and event data for the Packet Switched (PS) domain |
| TS 32.101 | Telecommunication management; Principles and high level requirements |
| TS 32.102 | Telecommunication management; Architecture |
| TS 32.103 | Telecommunication management; Integration Reference Point (IRP) overview and usage guide |
| TS 32.104 | Telecommunication management; 3G Performance Management |
| TS 32.106-1 | Telecommunication management; Configuration Management (CM); Part 1: Concept and requirements |
| TS 32.106-2 | Telecommunication management; Configuration Management (CM); Part 2: Notification Integration Reference Point (IRP): Information Service (IS) |

| | |
|---|---|
| TS 32.106-3 | Telecommunication management; Configuration Management (CM); Part 3: Notification Integration Reference Point (IRP); Common Object Request Broker Architecture (CORBA) Solution Set (SS) |
| TS 32.106-4 | Telecommunication management; Configuration Management (CM); Part 4: Notification Integration Reference Point (IRP); Common Management Information Protocol (CMIP) Solution Set (SS) |
| TS 32.106-5 | Telecommunication management; Configuration Management (CM); Part 5: Basic CM Integration Reference Point (IRP): Information model (including Network Resource Model (NRM) |
| TS 32.106-6 | Telecommunication management; Configuration Management (CM); Part 6: Basic CM Integration Reference Point (IRP): Common Object Request Broker Architecture (CORBA) Solution Set (SS) |
| TS 32.106-7 | Telecommunication management; Configuration Management (CM); Part 7: Basic CM Integration Reference Point (IRP): Common Management Information Protocol (CMIP) Solution Set (SS) |
| TS 32.106-8 | Telecommunication management; Configuration Management (CM); Part 8: Name convention for Managed Objects |
| TS 32.107 | Telecommunication management; Fixed Mobile Convergence (FMC) Federated Network Information Model (FNIM) |
| TS 32.111-1 | Telecommunication management; Fault Management; Part 1: 3G fault management requirements |
| TS 32.111-2 | Telecommunication management; Fault Management; Part 2: Alarm Integration Reference Point (IRP): Information Service (IS) |
| TS 32.111-3 | Telecommunication management; Fault Management; Part 3: Alarm Integration Reference Point (IRP): Common Object Request Broker Architecture (CORBA) Solution Set (SS) |
| TS 32.111-4 | Telecommunication management; Fault Management; Part 4: Alarm Integration Reference Point (IRP): Common Management Information Protocol (CMIP) Solution Set (SS) |

| | |
|---|---|
| TS 32.111-5 | Telecommunication management; Fault Management; Part 5: Alarm Integration Reference Point (IRP): eXtensible Markup Language (XML) definitions |
| TS 32.111-6 | Telecommunication management; Fault Management; Part 6: Alarm Integration Reference Point (IRP): Solution Set (SS) definitions |
| TS 32.111-7 | Telecommunication management; Fault Management; Part 7: Alarm IRP SOAP Solution Set (SS) |
| TS 32.121 | Telecommunication management; Advanced Alarm Management (AAM) Integration Reference Point (IRP): Requirements |
| TS 32.122 | Telecommunication management; Advanced Alarm Management (AAM) Integration Reference Point (IRP): Information Service (IS) |
| TS 32.123 | Telecommunication management; Advanced Alarm Management (AAM) Integration Reference Point (IRP); Common Object Request Broker Architecture (CORBA) Solution Set (SS) |
| TS 32.125 | Telecommunication management; Advanced Alarm Management (AAM) Integration Reference Point (IRP): eXtensible Markup Language (XML) file format definition |
| TS 32.126 | Telecommunication management; Advanced Alarm Management (AAM) Integration Reference Point (IRP); Solution Set (SS) definitions |
| TS 32.127 | Telecommunication management; Advanced Alarm Management (AAM) Integration Reference Point (IRP); SOAP Solution Set (SS) |
| TS 32.130 | Network sharing; Concepts and requirements |
| TS 32.140 | Telecommunication management; Subscription Management (SuM) requirements |
| TS 32.141 | Telecommunication management; Subscription Management (SuM) architecture |

| | |
|---|---|
| TS 32.150 | Telecommunication management; Integration Reference Point (IRP) Concept and definitions |
| TS 32.151 | Telecommunication management; Integration Reference Point (IRP) Information Service (IS) template |
| TS 32.152 | Telecommunication management; Integration Reference Point (IRP) Information Service (IS) Unified Modeling Language (UML) repertoire |
| TS 32.153 | Telecommunication management; Integration Reference Point (IRP) technology specific templates, rules and guidelines |
| TS 32.154 | Telecommunication management; Backward and Forward Compatibility (BFC); Concept and definitions |
| TS 32.155 | Telecommunication management; Requirements template |
| TS 32.156 | Telecommunication management; Fixed Mobile Convergence (FMC) model repertoire |
| TS 32.157 | Telecommunication management; Integration Reference Point (IRP) Information Service (IS) template |
| TS 32.171 | Telecommunication management; Subscription Management (SuM) Network Resource Model (NRM) Integration Reference Point (IRP); Requirements |
| TS 32.172 | Telecommunication management; Subscription Management (SuM) Network Resource Model (NRM) Integration Reference Point (IRP): Information Service (IS) |
| TS 32.175 | Telecommunication management; Subscription Management (SuM) Network Resource Model (NRM) Integration Reference Point (IRP): eXtensible Markup Language (XML) definition |
| TS 32.176 | Telecommunication management; Subscription Management (SuM) Network Resource Model (NRM) Integration Reference Point (IRP); Solution Set (SS) definitions |

| | |
|---|---|
| TS 32.181 | Telecommunication management; User Data Convergence (UDC); Framework for Model Handling and Management |
| TS 32.182 | Telecommunication management; User Data Convergence (UDC); Common baseline information model (CBIM) |
| TS 32.200 | Telecommunication management; Charging management; Charging principles |
| TS 32.205 | Telecommunication management; Charging management; Charging data description for the Circuit Switched (CS) domain |
| TS 32.215 | Telecommunication management; Charging management; Charging data description for the Packet Switched (PS) domain |
| TS 32.225 | Telecommunication management; Charging management; Charging data description for the IP Multimedia Subsystem (IMS) |
| TS 32.235 | Telecommunication management; Charging management; Charging data description for application services |
| TS 32.240 | Telecommunication management; Charging management; Charging architecture and principles |
| TS 32.250 | Telecommunication management; Charging management; Circuit Switched (CS) domain charging |
| TS 32.251 | Telecommunication management; Charging management; Packet Switched (PS) domain charging |
| TS 32.252 | Telecommunication management; Charging management; Wireless Local Area Network (WLAN) charging |
| TS 32.260 | Telecommunication management; Charging management; IP Multimedia Subsystem (IMS) charging |
| TS 32.270 | Telecommunication management; Charging management; Multimedia Messaging Service (MMS) charging |

| | |
|---|---|
| TS 32.271 | Telecommunication management; Charging management; Location Services (LCS) charging |
| TS 32.272 | Telecommunication management; Charging management; Push-to-talk over Cellular (PoC) charging |
| TS 32.273 | Telecommunication management; Charging management; Multimedia Broadcast and Multicast Service (MBMS) charging |
| TS 32.274 | Telecommunication management; Charging management; Short Message Service (SMS) charging |
| TS 32.275 | Telecommunication management; Charging management; MultiMedia Telephony (MMTel) charging |
| TS 32.276 | Telecommunication management; Charging management; Voice call service charging |
| TS 32.280 | Telecommunication management; Charging management; Advice of Charge (AoC) service |
| TS 32.293 | Telecommunication management; Charging management; Proxy function |
| TS 32.295 | Telecommunication management; Charging management; Charging Data Record (CDR) transfer |
| TS 32.296 | Telecommunication management; Charging management; Online Charging System (OCS): Applications and interfaces |
| TS 32.297 | Telecommunication management; Charging management; Charging Data Record (CDR) file format and transfer |
| TS 32.298 | Telecommunication management; Charging management; Charging Data Record (CDR) parameter description |
| TS 32.299 | Telecommunication management; Charging management; Diameter charging applications |

| | |
|---|---|
| TS 32.300 | Telecommunication management; Configuration Management (CM); Name convention for Managed Objects |
| TS 32.301 | Telecommunication management; Configuration Management (CM); Notification Integration Reference Point (IRP); Requirements |
| TS 32.302 | Telecommunication management; Configuration Management (CM); Notification Integration Reference Point (IRP); Information Service (IS) |
| TS 32.303 | Telecommunication management; Configuration Management (CM); Notification Integration Reference Point (IRP); Common Object Request Broker Architecture (CORBA) Solution Set (SS) |
| TS 32.304 | Telecommunication management; Configuration Management (CM); Notification Integration Reference Point (IRP): Common Management Information Protocol (CMIP) Solution Set (SS) |
| TS 32.305 | Telecommunication management; Configuration Management (CM); Notification Integration Reference Point (IRP); eXtensible Markup Language (XML) definition |
| TS 32.306 | Telecommunication management; Configuration Management (CM); Notification Integration Reference Point (IRP): Solution Set (SS) definitions |
| TS 32.307 | Telecommunication management; Configuration Management (CM); Notification Integration Reference Point (IRP); SOAP Solution Set (SS) |
| TS 32.311 | Telecommunication management; Generic Integration Reference Point (IRP) management; Requirements |
| TS 32.312 | Telecommunication management; Generic Integration Reference Point (IRP) management; Information Service (IS) |
| TS 32.313 | Telecommunication management; Generic Integration Reference Point (IRP) management; Common Object Request Broker Architecture (CORBA) Solution Set (SS) |

| | |
|---|---|
| TS 32.314 | Telecommunication management; Generic Integration Reference Point (IRP) management; Common Management Information Protocol (CMIP) Solution Set (SS) |
| TS 32.316 | Telecommunication management; Generic Integration Reference Point (IRP) management; Solution Set (SS) definitions |
| TS 32.317 | Telecommunication management; Generic Integration Reference Point (IRP) management; SOAP Solution Set (SS) |
| TS 32.321 | Telecommunication management; Test management Integration Reference Point (IRP); Requirements |
| TS 32.322 | Telecommunication management; Test management Integration Reference Point (IRP); Information Service (IS) |
| TS 32.323 | Telecommunication management; Test management Integration Reference Point (IRP); Common Object Request Broker Architecture (CORBA) Solution Set (SS) |
| TS 32.324 | Telecommunication management; Test management Integration Reference Point (IRP); Common Management Information Protocol (CMIP) Solution Set (SS) |
| TS 32.325 | Telecommunication management; Test management Integration Reference Point (IRP); eXtensible Markup Language (XML) definitions |
| TS 32.326 | Telecommunication management; Test management Integration Reference Point (IRP): Solution Set (SS) definitions |
| TS 32.327 | Telecommunication management; Test management Integration Reference Point (IRP); SOAP Solution Set (SS) |
| TS 32.331 | Telecommunication management; Notification Log (NL) Integration Reference Point (IRP); Requirements |
| TS 32.332 | Telecommunication management; Notification Log (NL) Integration Reference Point (IRP); Information Service (IS) |

| | |
|---|---|
| TS 32.333 | Telecommunication management; Notification Log (NL) Integration Reference Point (IRP); Common Object Request Broker Architecture (CORBA) Solution Set (SS) |
| TS 32.334 | Telecommunication management; Notification Log (NL) Integration Reference Point (IRP); Common Management Information Protocol (CMIP) Solution Set (SS) |
| TS 32.335 | Telecommunication management; Notification Log (NL) Integration Reference Point (IRP); eXtensible Markup Language (XML) solution definitions |
| TS 32.336 | Telecommunication management; Notification Log (NL) Integration Reference Point (IRP): Solution Set (SS) definitions |
| TS 32.337 | Telecommunication management; Notification Log (NL) Integration Reference Point (IRP); SOAP Solution Set (SS) |
| TS 32.341 | Telecommunication management; File Transfer (FT) Integration Reference Point (IRP); Requirements |
| TS 32.342 | Telecommunication management; File Transfer (FT) Integration Reference Point (IRP); Information Service (IS) |
| TS 32.343 | Telecommunication management; File Transfer (FT) Integration Reference Point (IRP); Common Object Request Broker Architecture (CORBA) Solution Set (SS) |
| TS 32.344 | Telecommunication management; File Transfer (FT) Integration Reference Point (IRP); Common Management Information Protocol (CMIP) Solution Set (SS) |
| TS 32.345 | Telecommunication management; File Transfer (FT) Integration Reference Point (IRP); eXtensible Markup Language (XML) definitions |
| TS 32.346 | Telecommunication management; File Transfer (FT) Integration Reference Point (IRP): Solution Set (SS) definitions |

| | |
|---|---|
| TS 32.347 | Telecommunication management; File Transfer (FT) Integration Reference Point (IRP); SOAP Solution Set (SS) |
| TS 32.351 | Telecommunication management; Communication Surveillance (CS) Integration Reference Point (IRP); Requirements |
| TS 32.352 | Telecommunication management; Communication Surveillance (CS) Integration Reference Point (IRP); Information Service (IS) |
| TS 32.353 | Telecommunication management; Communication Surveillance (CS) Integration Reference Point (IRP); Common Object Request Broker Architecture (CORBA) Solution Set (SS) |
| TS 32.354 | Telecommunication management; Communication Surveillance (CS) Integration Reference Point (IRP); Common Management Information Protocol (CMIP) Solution Set (SS) |
| TS 32.355 | Telecommunication management; Communication Surveillance (CS) Integration Reference Point (IRP); eXtensible Markup Language (XML) definitions |
| TS 32.356 | Telecommunication management; Communication Surveillance (CS) Integration Reference Point (IRP); Solution Set (SS) definitions |
| TS 32.357 | Telecommunication management; Communication Surveillance (CS) Integration Reference Point (IRP); SOAP Solution Set (SS) |
| TS 32.361 | Telecommunication management; Entry Point (EP) Integration Reference Point (IRP); Requirements |
| TS 32.362 | Telecommunication management; Entry Point (EP) Integration Reference Point (IRP); Information Service (IS) |
| TS 32.363 | Telecommunication management; Entry Point (EP) Integration Reference Point (IRP); Common Object Request Broker Architecture (CORBA) Solution Set (SS) |

| | |
|---|---|
| TS 32.364 | Telecommunication Management; Entry Point (EP) Integration Reference Point (IRP); Common Management Information Protocol (CMIP) Solution Set (SS) |
| TS 32.365 | Telecommunication management; Entry Point (EP) Integration Reference Point (IRP); eXtensible Markup Language (XML) definitions |
| TS 32.366 | Telecommunication management; Entry Point (EP) Integration Reference Point (IRP); Solution Set (SS) definitions |
| TS 32.367 | Telecommunication management; Entry Point (EP) Integration Reference Point (IRP); SOAP Solution Set (SS) |
| TS 32.371 | Telecommunication management; Security Management concept and requirements |
| TS 32.372 | Telecommunication management; Security services for Integration Reference Point (IRP); Information Service (IS) |
| TS 32.373 | Telecommunication management; Security services for Integration Reference Point (IRP); Common Object Request Broker Architecture (CORBA) solution |
| TS 32.375 | Telecommunication management; Security services for Integration Reference Point (IRP); File integrity solution |
| TS 32.376 | Telecommunication management; Security services for Integration Reference Point (IRP); Solution Set (SS) definitions |
| TS 32.381 | Telecommunication management; Partial Suspension of Itf-N Integration Reference Point (IRP); Requirements |
| TS 32.382 | Telecommunication management; Partial Suspension of Itf-N Integration Reference Point (IRP); Information Service (IS) |
| TS 32.383 | Telecommunication management; Partial Suspension of Itf-N Integration Reference Point (IRP); Common Object Request Broker Architecture (CORBA) Solution Set (SS) |

| | |
|---|---|
| TS 32.385 | Telecommunication management; Partial Suspension of Itf-N Integration Reference Point (IRP): eXtensible Markup Language (XML) file format definition |
| TS 32.386 | Telecommunication management; Partial Suspension of Itf-N Integration Reference Point (IRP); Solution Set (SS) definitions |
| TS 32.387 | Telecommunication management; Partial Suspension of Itf-N Integration Reference Point (IRP); SOAP Solution Set (SS) |
| TS 32.391 | Telecommunication management; Delta synchronization Integration Reference Point (IRP); Requirements |
| TS 32.392 | Telecommunication management; Delta synchronization Integration Reference Point (IRP); Information Service (IS) |
| TS 32.393 | Telecommunication management; Delta synchronization Integration Reference Point (IRP); Common Object Request Broker Architecture (CORBA) Solution Set (SS) |
| TS 32.395 | Telecommunication management; Delta synchronization Integration Reference Point (IRP); eXtensible Markup Language (XML) file format definition |
| TS 32.396 | Telecommunication management; Delta synchronization Integration Reference Point (IRP); Solution Set (SS) definitions |
| TS 32.397 | Telecommunication management; Delta synchronization Integration Reference Point (IRP); SOAP Solution Set (SS) |
| TS 32.401 | Telecommunication management; Performance Management (PM); Concept and requirements |
| TS 32.403 | Telecommunication management; Performance Management (PM); Performance measurements; UMTS and combined UMTS/GSM |
| TS 32.404 | Telecommunication management; Performance Management (PM); Performance measurements; Definitions and template |

| | |
|---|---|
| TS 32.405 | Telecommunication management; Performance Management (PM); Performance measurements; Universal Terrestrial Radio Access Network (UTRAN) |
| TS 32.406 | Telecommunication management; Performance Management (PM); Performance measurements; Core Network (CN) Packet Switched (PS) domain |
| TS 32.407 | Telecommunication management; Performance Management (PM); Performance measurements; Core Network (CN) Circuit Switched (CS) domain; UMTS and combined UMTS/GSM |
| TS 32.408 | Telecommunication management; Performance Management (PM); Performance measurements; Teleservice |
| TS 32.409 | Telecommunication management; Performance Management (PM); Performance measurements; IP Multimedia Subsystem (IMS) |
| TS 32.410 | Telecommunication management; Key Performance Indicators (KPI) for UMTS and GSM |
| TS 32.411 | Telecommunication management; Performance Management (PM) Integration Reference Point (IRP): Requirements |
| TS 32.412 | Telecommunication management; Performance Management (PM) Integration Reference Point (IRP): Information Service (IS) |
| TS 32.413 | Telecommunication management; Performance Management (PM) Integration Reference Point (IRP): Common Object Request Broker Architecture (CORBA) Solution Set (SS) |
| TS 32.414 | Telecommunication management; Performance Management (PM) Integration Reference Point (IRP): Common Management Information Protocol (CMIP) Solution Set (SS) |
| TS 32.415 | Telecommunication management; Performance Management (PM) Integration Reference Point (IRP); eXtensible Markup Language (XML) definitions |

| | |
|---|---|
| TS 32.416 | Telecommunication management; Performance Management (PM) Integration Reference Point (IRP); Solution Set (SS) definitions |
| TS 32.417 | Telecommunication management; Performance Management (PM) Integration Reference Point (IRP); SOAP Solution Set (SS) |
| TS 32.421 | Telecommunication management; Subscriber and equipment trace; Trace concepts and requirements |
| TS 32.422 | Telecommunication management; Subscriber and equipment trace; Trace control and configuration management |
| TS 32.423 | Telecommunication management; Subscriber and equipment trace; Trace data definition and management |
| TS 32.425 | Telecommunication management; Performance Management (PM); Performance measurements Evolved Universal Terrestrial Radio Access Network (E-UTRAN) |
| TS 32.426 | Telecommunication management; Performance Management (PM); Performance measurements Evolved Packet Core (EPC) network |
| TS 32.432 | Telecommunication management; Performance measurement: File format definition |
| TS 32.435 | Telecommunication management; Performance measurement; eXtensible Markup Language (XML) file format definition |
| TS 32.436 | Telecommunication management; Performance measurement: Abstract Syntax Notation 1 (ASN.1) file format definition |
| TS 32.441 | Telecommunication management; Trace Management Integration Reference Point (IRP); Requirements |
| TS 32.442 | Telecommunication management; Trace Management Integration Reference Point (IRP); Information Service (IS) |

| | |
|---|---|
| TS 32.443 | Telecommunication management; Trace Management (Trace) Integration Reference Point (IRP): Common Object Request Broker Architecture (CORBA) Solution Set (SS) |
| TS 32.445 | Telecommunication management; Trace Management (Trace) Integration Reference Point (IRP): eXtensible Markup Language (XML) file format definition |
| TS 32.446 | Telecommunication management; Trace Management Integration Reference Point (IRP); Solution Set (SS) definitions |
| TS 32.447 | Telecommunication management; Trace Management Integration Reference Point (IRP): SOAP Solution Set (SS) |
| TS 32.450 | Telecommunication management; Key Performance Indicators (KPI) for Evolved Universal Terrestrial Radio Access Network (E-UTRAN): Definitions |
| TS 32.451 | Telecommunication management; Key Performance Indicators (KPI) for Evolved Universal Terrestrial Radio Access Network (E-UTRAN); Requirements |
| TS 32.452 | Telecommunication management; Performance Management (PM); Performance measurements Home Node B (HNB) Subsystem (HNS) |
| TS 32.453 | Telecommunication management; Performance Management (PM); Performance measurements Home enhanced Node B (HeNB) Subsystem (HeNS) |
| TS 32.454 | Telecommunication management; Key Performance Indicators (KPI) for the IP Multimedia Subsystem (IMS); Definitions |
| TS 32.455 | Telecommunication management; Key Performance Indicators (KPI) for the Evolved Packet Core (EPC); Definitions |
| TS 32.500 | Telecommunication management; Self-Organizing Networks (SON); Concepts and requirements |

| | |
|---|---|
| TS 32.501 | Telecommunication management; Self-configuration of network elements; Concepts and requirements |
| TS 32.502 | Telecommunication management; Self-configuration of network elements Integration Reference Point (IRP); Information Service (IS) |
| TS 32.503 | Telecommunication management; Self-configuration of network elements Integration Reference Point (IRP); Common Object Request Broker Architecture (CORBA) Solution Set (SS) |
| TS 32.505 | Telecommunication management; Self-configuration of network elements Integration Reference Point (IRP): eXtensible Markup Language (XML) file format definition |
| TS 32.506 | Telecommunication management; Self-configuration of network elements Integration Reference Point (IRP); Solution Set (SS) definitions |
| TS 32.507 | Telecommunication management; Self-configuration of network elements Integration Reference Point (IRP): SOAP Solution Set (SS) |
| TS 32.508 | Telecommunication management; Procedure flows for multi-vendor plug-and-play eNode B connection to the network |
| TS 32.509 | Telecommunication management; Data formats for multi-vendor plug and play eNode B connection to the network |
| TS 32.511 | Telecommunication management; Automatic Neighbor Relation (ANR) management; Concepts and requirements |
| TS 32.521 | Telecommunication management; Self-Organizing Networks (SON) Policy Network Resource Model (NRM) Integration Reference Point (IRP); Requirements |
| TS 32.522 | Telecommunication management; Self-Organizing Networks (SON) Policy Network Resource Model (NRM) Integration Reference Point (IRP); Information Service (IS) |

| | |
|---|---|
| TS 32.523 | Telecommunication management; Self-Organizing Networks (SON); Policy Network Resource Model (NRM) Integration Reference Point (IRP); Common Object Request Broker Architecture (CORBA) Solution Set (SS) |
| TS 32.525 | Telecommunication management; Self-Organizing Networks (SON) Policy Network Resource Model (NRM) Integration Reference Point (IRP); eXtensible Markup Language (XML) file format definition |
| TS 32.526 | Telecommunication management; Self-Organizing Networks (SON); Policy Network Resource Model (NRM) Integration Reference Point (IRP); Solution Set (SS) definitions |
| TS 32.531 | Telecommunication management; Software management (SwM); Concepts and Integration Reference Point (IRP) Requirements |
| TS 32.532 | Telecommunication management; Software management (SwM); Integration Reference Point (IRP); Information Service (IS) |
| TS 32.533 | Telecommunication management; Software management (SwM); Integration Reference Point (IRP); Common Object Request Broker Architecture (CORBA) Solution Set (SS) |
| TS 32.535 | Telecommunication management; Software Management Integration Reference Point (IRP); eXtensible Markup Language (XML) definitions |
| TS 32.536 | Telecommunication management; Software management Integration Reference Point (IRP); Solution Set (SS) definitions |
| TS 32.537 | Telecommunication management; Software management (SwM); Integration Reference Point (IRP); SOAP Solution Set (SS) |
| TS 32.541 | Telecommunication management; Self-Organizing Networks (SON); Self-healing concepts and requirements |
| TS 32.551 | Telecommunication management; Energy Saving Management (ESM); Concepts and requirements |

| | |
|---|---|
| TS 32.571 | Telecommunication management; Home Node B (HNB) and Home eNode B (HeNB) management; Type 2 interface concepts and requirements |
| TS 32.572 | Telecommunication management; Home Node B (HNB) and Home eNode B (HeNB) management; Type 2 interface models and mapping functions |
| TS 32.581 | Telecommunication management; Home Node B (HNB) Operations, Administration, Maintenance and Provisioning (OAM&P); Concepts and requirements for Type 1 interface HNB to HNB Management System (HMS) |
| TS 32.582 | Telecommunication management; Home Node B (HNB) Operations, Administration, Maintenance and Provisioning (OAM&P); Information model for Type 1 interface HNB to HNB Management System (HMS) |
| TS 32.583 | Telecommunication management; Home Node B (HNB) Operations, Administration, Maintenance and Provisioning (OAM&P); Procedure flows for Type 1 interface HNB to HNB Management System (HMS) |
| TS 32.584 | Telecommunication management; Home Node B (HNB) Operations, Administration, Maintenance and Provisioning (OAM&P); XML definitions for Type 1 interface HNB to HNB Management System (HMS) |
| TS 32.591 | Telecommunication management; Home enhanced Node B (HeNB) Operations, Administration, Maintenance and Provisioning (OAM&P); Concepts and requirements for Type 1 interface HeNB to HeNB Management System (HeMS) |
| TS 32.592 | Telecommunication management; Home enhanced Node B (HeNB) Operations, Administration, Maintenance and Provisioning (OAM&P); Information model for Type 1 interface HeNB to HeNB Management System (HeMS) |

| | |
|---|---|
| TS 32.593 | Telecommunication management; Home enhanced Node B (HeNB) Operations, Administration, Maintenance and Provisioning (OAM&P); Procedure flows for Type 1 interface HeNB to HeNB Management System (HeMS) |
| TS 32.594 | Telecommunication management; Home enhanced Node B (HeNB) Operations, Administration, Maintenance and Provisioning (OAM&P); XML definitions for Type 1 interface HeNB to HeNB Management System (HeMS) |
| TS 32.600 | Telecommunication management; Configuration Management (CM); Concept and high-level requirements |
| TS 32.601 | Telecommunication management; Configuration Management (CM); Basic CM Integration Reference Point (IRP); Requirements |
| TS 32.602 | Telecommunication management; Configuration Management (CM); Basic CM Integration Reference Point (IRP); Information Service (IS) |
| TS 32.603 | Telecommunication management; Configuration Management (CM); Basic CM Integration Reference Point (IRP); Common Object Request Broker Architecture (CORBA) Solution Set (SS) |
| TS 32.604 | Telecommunication management; Configuration Management (CM); Basic CM Integration Reference Point (IRP) Common Management Information Protocol (CMIP) Solution Set (SS) |
| TS 32.606 | Telecommunication management; Configuration Management (CM); Basic CM Integration Reference Point (IRP); Solution Set (SS) definitions |
| TS 32.607 | Telecommunication management; Configuration Management (CM); Basic CM Integration Reference Point (IRP); SOAP Solution Set (SS) |
| TS 32.611 | Telecommunication management; Configuration Management (CM); Bulk CM Integration Reference Point (IRP): Requirements |

| | |
|---|---|
| TS 32.612 | Telecommunication management; Configuration Management (CM); Bulk CM Integration Reference Point (IRP): Information Service (IS) |
| TS 32.613 | Telecommunication management; Configuration Management (CM); Bulk CM Integration Reference Point (IRP): Common Object Request Broker Architecture (CORBA) Solution Set (SS) |
| TS 32.614 | Telecommunication management; Configuration Management (CM); Bulk CM Integration Reference Point (IRP): Common Management Information Protocol (CMIP) Solution Set (SS) |
| TS 32.615 | Telecommunication management; Configuration Management (CM); Bulk CM Integration Reference Point (IRP): eXtensible Markup Language (XML) file format definition |
| TS 32.616 | Telecommunication management; Configuration Management (CM); Bulk CM Integration Reference Point (IRP); Solution Set (SS) definitions |
| TS 32.617 | Telecommunication management; Configuration Management (CM); Bulk CM Integration Reference Point (IRP): Bulk CM IRP SOAP Solution Set (SS) |
| TS 32.621 | Telecommunication management; Configuration Management (CM); Generic network resources Integration Reference Point (IRP); Requirements |
| TS 32.622 | Telecommunication management; Configuration Management (CM); Generic network resources Integration Reference Point (IRP); Network Resource Model (NRM) |
| TS 32.623 | Telecommunication management; Configuration Management (CM); Generic network resources Integration Reference Point (IRP); Common Object Request Broker Architecture (CORBA) Solution Set (SS) |
| TS 32.624 | Telecommunication management; Configuration Management (CM); Generic network resources: Integration Reference Point (IRP): Common Management Information Protocol (CMIP) Solution Set (SS) |

| | |
|---|---|
| TS 32.625 | Telecommunication management; Configuration Management (CM); Generic network resources Integration Reference Point (IRP): Bulk CM eXtensible Markup Language (XML) file format definition |
| TS 32.626 | Telecommunication management; Configuration Management (CM); Generic network resources Integration Reference Point (IRP); Solution Set (SS) definitions |
| TS 32.631 | Telecommunication management; Configuration Management (CM); Core network resources Integration Reference Point (IRP); Requirements |
| TS 32.632 | Telecommunication management; Configuration Management (CM); Core Network Resources Integration Reference Point (IRP); Network Resource Model (NRM) |
| TS 32.633 | Telecommunication management; Configuration Management (CM); Core network resources Integration Reference Point (IRP); Common Object Request Broker Architecture (CORBA) Solution Set (SS) |
| TS 32.634 | Telecommunication management; Configuration Management (CM); Core network resources Integration Reference Point (IRP): Common Management Information Protocol (CMIP) Solution Set (SS) |
| TS 32.635 | Telecommunication management; Configuration Management (CM); Core network resources Integration Reference Point (IRP); Bulk CM eXtensible Markup Language (XML) file format definition |
| TS 32.636 | Telecommunication management; Configuration Management (CM); Core network resources Integration Reference Point (IRP); Solution Set (SS) Definitions |
| TS 32.641 | Telecommunication management; Configuration Management (CM); UTRAN network resources Integration Reference Point (IRP); Requirements |
| TS 32.642 | Telecommunication management; Configuration Management (CM); UTRAN network resources Integration Reference Point (IRP); Network Resource Model (NRM) |

| | |
|---|---|
| TS 32.643 | Telecommunication management; Configuration Management (CM); UTRAN network resources Integration Reference Point (IRP); Common Object Request Broker Architecture (CORBA) Solution Set (SS) |
| TS 32.644 | Telecommunication management; Configuration Management (CM); UTRAN network resources Integration Reference Point (IRP): Common Management Information Protocol (CMIP) Solution Set (SS) |
| TS 32.645 | Telecommunication management; Configuration Management (CM); UTRAN network resources Integration Reference Point (IRP); Bulk CM eXtensible Markup Language (XML) file format definition |
| TS 32.646 | Telecommunication management; Configuration Management (CM); UTRAN network resources Integration Reference Point (IRP); Solution Set (SS) definitions |
| TS 32.651 | Telecommunication management; Configuration Management (CM); GERAN network resources Integration Reference Point (IRP); Requirements |
| TS 32.652 | Telecommunication management; Configuration Management (CM); GERAN network resources Integration Reference Point (IRP); Network Resource Model (NRM) |
| TS 32.653 | Telecommunication management; Configuration Management (CM); GERAN network resources Integration Reference Point (IRP); Common Object Request Broker Architecture (CORBA) Solution Set (SS) |
| TS 32.654 | Telecommunication management; Configuration Management (CM); GERAN network resources Integration Reference Point (IRP): Common Management Information Protocol (CMIP) Solution Set (SS) |
| TS 32.655 | Telecommunication management; Configuration Management (CM); GERAN network resources Integration Reference Point (IRP); Bulk CM eXtensible Markup Language (XML) file format definition |

| | |
|---|---|
| TS 32.656 | Telecommunication management; Configuration Management (CM); GERAN network resources Integration Reference Point (IRP); Solution Set (SS) definitions |
| TS 32.661 | Telecommunication management; Configuration Management (CM); Kernel CM Requirements |
| TS 32.662 | Telecommunication management; Configuration Management (CM); Kernel CM Information Service (IS) |
| TS 32.663 | Telecommunication management; Configuration Management (CM); Kernel CM Integration Reference Point (IRP); Common Object Request Broker Architecture (CORBA) Solution Set (SS) |
| TS 32.664 | Telecommunication management; Configuration Management (CM); Kernel CM Integration Reference Point (IRP); Common Management Information Protocol (CMIP) Solution Set (SS) |
| TS 32.665 | Telecommunication management; Configuration Management (CM); Kernel CM Integration Reference Point (IRP); eXtensible Markup Language (XML) definitions |
| TS 32.666 | Telecommunication management; Configuration Management (CM); Kernel CM Integration Reference Point (IRP); Solution Set (SS) definitions |
| TS 32.667 | Telecommunication management; Configuration Management (CM); Kernel CM Integration Reference Point (IRP); SOAP Solution Set (SS) |
| TS 32.671 | Telecommunication management; Configuration Management (CM); State Management Integration Reference Point (IRP); Requirements |
| TS 32.672 | Telecommunication management; Configuration Management (CM); State Management Integration Reference Point (IRP); Information Service (IS) |
| TS 32.673 | Telecommunication management; Configuration Management (CM); State Management Integration Reference Point (IRP); Common Object Request Broker Architecture (CORBA) Solution Set (SS) |

| | |
|---|---|
| TS 32.674 | Telecommunication management; Configuration Management (CM); State Management Integration Reference Point (IRP): Common Management Information Protocol (CMIP) Solution Set (SS) |
| TS 32.675 | Telecommunication management; Configuration Management (CM); State Management Integration Reference Point (IRP); Bulk CM eXtensible Markup Language (XML) file format definition |
| TS 32.676 | Telecommunication management; Configuration Management (CM); State Management Integration Reference Point (IRP); Solution Set (SS) definitions |
| TS 32.690 | Telecommunication management; Inventory Management (IM); Requirements |
| TS 32.691 | Telecommunication management; Inventory Management (IM) network resources Integration Reference Point (IRP): Requirements |
| TS 32.692 | Telecommunication management; Inventory Management (IM) network resources Integration Reference Point (IRP); Network Resource Model (NRM) |
| TS 32.695 | Telecommunication management; Inventory Management (IM) network resources Integration Reference Point (IRP); Bulk Configuration Management (CM) eXtensible Markup Language (XML) file format definition |
| TS 32.696 | Telecommunication management; Inventory Management (IM) Network Resource Model (NRM) Integration Reference Point (IRP); Solution Set (SS) definitions |
| TS 32.711 | Telecommunication management; Configuration Management (CM); Transport Network (TN) interface Network Resource Model (NRM) Integration Reference Point (IRP); Requirements |
| TS 32.712 | Telecommunication management; Configuration Management (CM); Transport Network (TN) interface Network Resource Model (NRM) Integration Reference Point (IRP); Information Service (IS) |

| | |
|---|---|
| TS 32.713 | Telecommunication management; Configuration Management (CM); Transport Network (TN) Network Resource Model (NRM) Integration Reference Point (IRP); Common Object Request Broker Architecture (CORBA) Solution Set (SS) |
| TS 32.714 | Telecommunication management; Configuration Management (CM); Transport Network (TN) Network Resource Model (NRM) Integration Reference Point (IRP); Common Management Information Protocol (CMIP) Solution Set (SS) |
| TS 32.715 | Telecommunication management; Configuration Management (CM) Transport Network (TN); Network Resource Model (NRM) Integration Reference Point (IRP); Bulk CM eXtensible Markup Language (XML) file format definition |
| TS 32.716 | Telecommunication management; Configuration Management (CM); Transport Network (TN) interface Network Resource Model (NRM) Integration Reference Point (IRP); Solution Set (SS) definitions |
| TS 32.721 | Telecommunication management; Configuration Management (CM); Repeater network resources Integration Reference Point (IRP); Requirements |
| TS 32.722 | Telecommunication management; Configuration Management (CM); Repeater network resources Integration Reference Point (IRP); information Service (IS) |
| TS 32.723 | Telecommunication management; Configuration Management (CM); Repeater network resources Integration Reference Point (IRP); Common Object Request Broker Architecture (CORBA) Solution Set (SS) |
| TS 32.725 | Telecommunication management; Configuration Management (CM); Repeater network resources Integration Reference Point (IRP); Bulk CM eXtensible Markup Language (XML) file format definition |
| TS 32.726 | Telecommunication management; Configuration Management (CM); Repeater network resources Integration Reference Point (IRP); Solution Set (SS) definitions |

| | |
|---|---|
| TS 32.731 | Telecommunication management; IP Multimedia Subsystem (IMS) Network Resource Model (NRM) Integration Reference Point (IRP); Requirements |
| TS 32.732 | Telecommunication management; IP Multimedia Subsystem (IMS) Network Resource Model (NRM) Integration Reference Point (IRP); Information Service (IS) |
| TS 32.733 | Telecommunication management; IP Multimedia Subsystem (IMS) Network Resource Model (NRM) Integration Reference Point (IRP); Common Object Request Broker Architecture (CORBA) Solution Set (SS) |
| TS 32.735 | Telecommunication management; IP Multimedia Subsystem (IMS) Network Resource Model (NRM) Integration Reference Point (IRP); Bulk CM eXtensible Markup Language (XML) file format definition |
| TS 32.736 | Telecommunication management; IP Multimedia Subsystem (IMS) Network Resource Model (NRM) Integration Reference Point (IRP); Solution Set (SS) definitions |
| TS 32.741 | Telecommunication management; Configuration Management (CM); Signalling Transport Network (STN) interface Network Resource Model (NRM) Integration Reference Point (IRP); Requirements |
| TS 32.742 | Telecommunication management; Configuration Management (CM); Signalling Transport Network (STN) interface Network Resource Model (NRM) Integration Reference Point (IRP); Information Service (IS) |
| TS 32.743 | Telecommunication management; Configuration Management (CM); Signalling Transport Network (STN) interface Network Resource Model (NRM) Integration Reference Point (IRP); Common Object Request Broker Architecture (CORBA) Solution Set (SS) |
| TS 32.744 | Telecommunication management; Configuration Management (CM); Signalling Transport Network (STN) interface Network Resource Model (NRM) Integration Reference Point (IRP): Common Management Information Protocol (CMIP) Solution Set (SS) |

| | |
|---|---|
| TS 32.745 | Telecommunication management; Configuration Management (CM); Signalling Transport Network (STN) interface Network Resource Model (NRM) Integration Reference Point (IRP); Bulk CM eXtensible Markup Language (XML) file format definition |
| TS 32.746 | Telecommunication management; Configuration Management (CM); Signalling Transport Network (STN) interface Network Resource Model (NRM) Integration Reference Point (IRP); Solution Set (SS) definitions |
| TS 32.751 | Telecommunication management; Evolved Packet Core (EPC) Network Resource Model (NRM) Integration Reference Point (IRP); Requirements |
| TS 32.752 | Telecommunication management; Evolved Packet Core (EPC) Network Resource Model (NRM) Integration Reference Point (IRP); Information Service (IS) |
| TS 32.753 | Telecommunication management; Evolved Packet Core (EPC) Network Resource Model (NRM) Integration Reference Point (IRP); Common Object Request Broker Architecture (CORBA) Solution Set (SS) |
| TS 32.755 | Telecommunication management; Evolved Packet Core (EPC) Network Resource Model (NRM) Integration Reference Point (IRP); eXtensible Markup Language (XML) definitions |
| TS 32.756 | Telecommunication management; Evolved Packet Core (EPC) Network Resource Model (NRM) Integration Reference Point (IRP); Solution Set (SS) Definitions |
| TS 32.761 | Telecommunication management; Evolved Universal Terrestrial Radio Access Network (E-UTRAN) Network Resource Model (NRM) Integration Reference Point (IRP); Requirements |
| TS 32.762 | Telecommunication management; Evolved Universal Terrestrial Radio Access Network (E-UTRAN) Network Resource Model (NRM) Integration Reference Point (IRP); Information Service (IS) |

| | |
|---|---|
| TS 32.763 | Telecommunication management; Evolved Universal Terrestrial Radio Access Network (E-UTRAN) Network Resource Model (NRM) Integration Reference Point (IRP): Common Object Request Broker Architecture (CORBA) Solution Set (SS) |
| TS 32.765 | Telecommunication management; Evolved Universal Terrestrial Radio Access Network (E-UTRAN) Network Resource Model (NRM) Integration Reference Point (IRP); eXtensible Markup Language (XML) definitions |
| TS 32.766 | Telecommunication management; Evolved Universal Terrestrial Radio Access Network (E-UTRAN) Network Resource Model (NRM) Integration Reference Point (IRP); Solution Set (SS) definitions |
| TS 32.771 | Telecommunication management; Home Node B (HNB) Subsystem (HNS); Network Resource Model (NRM); Integration Reference Point (IRP); Requirements |
| TS 32.772 | Telecommunication management; Home Node B (HNB) Subsystem (HNS); Network Resource Model (NRM); Integration Reference Point (IRP); Information Service (IS) |
| TS 32.773 | Telecommunication management; Home Node B (HNB) Subsystem (HNS); Network Resource Model (NRM); Integration Reference Point (IRP); Common Object Request Broker Architecture (CORBA) Solution Set (SS) |
| TS 32.775 | Telecommunication management; Home Node B (HNB) Subsystem (HNS); Network Resource Model (NRM); Integration Reference Point (IRP); eXtensible Markup Language (XML) file format definition |
| TS 32.776 | Telecommunication management; Home Node B (HNB) Subsystem (HNS); Network Resource Model (NRM); Integration Reference Point (IRP); Solution Set (SS) definitions |
| TS 32.781 | Telecommunication management; Home enhanced Node B (HeNB) Subsystem (HeNS); Network Resource Model (NRM); Integration Reference Point (IRP); Requirements |

| | |
|---|---|
| TS 32.782 | Telecommunication management; Home enhanced Node B (HeNB) Subsystem (HeNS); Network Resource Model (NRM); Integration Reference Point (IRP): Information Service (IS) |
| TS 32.783 | Telecommunication management; Home enhanced Node B (HeNB) Subsystem (HeNS); Network Resource Model (NRM); Integration Reference Point (IRP); Common Object Request Broker Architecture (CORBA) Solution Set (SS) |
| TS 32.785 | Telecommunication management; Home enhanced Node B (HeNB) Subsystem (HeNS); Network Resource Model (NRM); Integration Reference Point (IRP); Bulk CM eXtensible Markup Language (XML) file format definition |
| TS 32.786 | Telecommunication management; Home enhanced Node B (HeNB) Subsystem (HeNS); Network Resource Model (NRM); Integration Reference Point (IRP); Solution Set (SS) definitions |
| TS 32.791 | Telecommunication management; Common Radio Access Technology (RAT) Network Resource Model (NRM) Integration Reference Point (IRP); Requirements |
| TS 32.792 | Telecommunication management; Generic Radio Access Network (RAN) Network Resource Model (NRM) Integration Reference Point (IRP); Information Service (IS) |
| TS 32.796 | Telecommunication management; Generic Radio Access Network (RAN) Network Resource Model (NRM) Integration Reference Point (IRP); Solution Set (SS) definitions |
| TR 32.800 | Telecommunication management; Management level procedures and interaction with UTRAN |
| TR 32.802 | Telecommunication management; User Equipment Management (UEM) feasibility study |
| TR 32.803 | Telecommunication management; Process guide; Use cases in Unified Modeling Language (UML) |

| | |
|---|---|
| TR 32.804 | Telecommunication management; Control of Remote Electrical Tilting (RET) antennas; Requirements |
| TR 32.805 | Telecommunication management; Process guide; Backward compatibility recommendations |
| TR 32.806 | Telecommunication management; Application guide for use of Integration Reference Points (IRPs) on peer-to-peer (p2p) interface |
| TR 32.808 | Telecommunication management; Study of Common Profile Storage (CPS) Framework of User Data for network services and management |
| TR 32.809 | Telecommunication management; Feasibility study of XML-based (SOAP/HTTP) IRP solution sets |
| TR 32.810 | Telecommunication management; Information model Integration Reference Point (IRP) Study |
| TR 32.811 | Telecommunication management; Itf-N performance criteria: Requirements |
| TR 32.812 | Telecommunication management; Itf-N Implementation Conformance Statement (ICS) template |
| TR 32.814 | Telecommunication management; UTRAN and GERAN Key Performance Indicators (KPI) |
| TR 32.815 | Telecommunication management; Charging management; Online Charging System (OCS) architecture study |
| TR 32.816 | Telecommunication management; Study on management of Evolved Universal Terrestrial Radio Access Network (E-UTRAN) and Evolved Packet Core (EPC) |
| TR 32.818 | Telecommunication management; Study on 3GPP SA5 / MTOSI XML harmonization |

| | |
|---|---|
| TR 32.819 | Telecommunications management; Element management layer - Operation System Function (E-OSF) definition |
| TR 32.820 | Telecommunication management; Study on charging management; 3GPP Evolved Packet Core (EPC): Charging aspects |
| TR 32.821 | Telecommunication management; Study of Self-Organizing Networks (SON) related Operations, Administration and Maintenance (OAM) for Home Node B (HNB) |
| TR 32.822 | Telecommunication management; Study on System Maintenance over Itf-N |
| TR 32.823 | Telecommunication management; Self-Organizing Networks (SON); Study on self-healing |
| TR 32.824 | Telecommunication management; Service Oriented Architecture (SOA) Integration Reference Point (IRP) study |
| TR 32.825 | Telecommunication management; Charging management; Rc reference point study |
| TR 32.826 | Telecommunication management; Study on Energy Savings Management (ESM) |
| TR 32.827 | Telecommunication management; Integration of device management information with Itf-N |
| TR 32.828 | Telecommunication management; Study on alignment of 3GPP generic Network Resource Model (NRM) Integration Reference Point (IRP) and the TeleManagement Forum (TMF) Shared Information/Data (SID) model |
| TR 32.829 | Telecommunication management; Study on alignment of 3GPP alarm Integration Reference Point (IRP) and TeleManagement Forum (TMF) Interface Program (TIP) fault management |
| TR 32.830 | Telecommunication management; Study on version handling over Itf-N |

| | |
|---|---|
| TR 32.831 | Telecommunication management; Study on alignment of 3GPP Performance Management (PM) and TeleManagement Forum (TMF) Interface Program (TIP) Performance Management |
| TR 32.832 | Telecommunication management; Study on alarm correlation and alarm root cause analysis |
| TR 32.833 | Telecommunication management; Study on management of converged networks |
| TR 32.834 | Study on Operations, Administration and Maintenance (OAM) aspects of inter-Radio-Access-Technology (RAT) energy saving |
| TR 32.835 | Study on management of Heterogeneous Networks |
| TR 32.836 | Telecommunication management; Study on Network Management (NM) centralized Coverage and Capacity Optimization (CCO) Self-Organizing Networks (SON) function |
| TR 32.837 | Compliance of 3GPP OAM specifications to NGMN NGCOR |
| TR 32.838 | Study on Compliance of 3GPP OAM specifications to the NGMN Top OPE Recommendations |
| TR 32.839 | Management Specification Tooling Environment |
| TR 32.840 | Management Specification Testing Recommendations |
| TR 32.841 | WLAN impacts on Type-2 management |
| TR 32.849 | Study on Charging aspects of Roaming End-to-end scenarios with VoLTE IMS and interconnecting networks |
| TR 32.851 | Telecommunication management; Study on Operations, Administration and Maintenance (OAM) aspects of Network Sharing |
| TR 32.853 | Telecommunication management; Final report from the 3GPP – TM Forum Joint Work Group (JWG) on fault management harmonization |

| | |
|---|---|
| TR 32.854 | Telecommunication management; Fixed Mobile Convergence (FMC) 3GPP / TM Forum concrete model relationships and use cases |
| TR 32.858 | Usage Monitoring Control OCS enhancement |
| TR 32.859 | Telecommunication management; Study on Alarm Management |
| TR 32.860 | Telecommunication management; Study on enhancement of OAM aspects of distributed Self-Organizing Network (SON) functions |
| TR 32.861 | Telecommunication management; Study on Application and Partitioning of Itf-N |
| TR 32.901 | Telecommunication management; Study on User Data Convergence (UDC) information model handling and provisioning: Example use cases |
| TS 52.008 | Telecommunication management; GSM subscriber and equipment trace |
| TS 52.402 | Telecommunication management; Performance Management (PM); Performance measurements – GSM |

# 3

# OTT Services

## 3.1. Introduction

Over-the-top (OTT) for an economic actor is the position where it will take advantage of the investments of other actors for free and establish a fruitful business by offering its services to the customers of these actors. More generally, an OTT business can be anything that disrupts traditional billing models.

For a long time, such situations were impossible to create in the telecommunications and broadcasting business. Telecommunications were generally state-owned monopolies with very strict regulations. It was the same for television as long as the broadcasting means were limited and regulated. The only example was road traffic with heavy lorries traveling free throughout Europe without paying for the damage they caused to the roads.

### 3.1.1. *Impact of the technology*

The breakthrough of technology has changed the paradigm. Over several years, the Internet has been invented, as well as the Asymmetric Digital Subscriber Line (ADSL). Optical fibers have been laid, especially linking continents with immense capacities. The invention of wavelength division multiplex (WDM), followed by dense wavelength division multiplex (DWDM), put on the market unexpected transmission capacities. The bitrates carried on one fiber have been multiplied by more than 100. Telecommunications operators could not avoid selling these capacities at a very low price. Other breakthroughs have had an impact on television compression, with the advent of MPEG2, followed by on H264/MPEG4, and now H265/High Efficiency Video Coding (HEVC). Digital television with high definition (720p or 1080i) needed 1.5 Gbps without compression, with MPEG2 that was down to roughly 10 Mbps; MPEG4 lowers the bitrate to 5 Mbps, and H265 is expected to have it with less than 3 Mbps. These technological achievements open the way to Web television. It is no longer necessary to pay for the network operator's managed Internet Protocol Television (IPTV); with the help of some clever engineering, the same service is offered only as a flow of packets. Even telephony can be offered on an OTT base.

Thus, OTT is mainly developed in telecommunications and video services. There is a strong and solvent demand for receiving music and/or movies with a large choice of productions. With a different financing, Internet actors from the USA could overwhelm the world with pseudo-free services while they collect fees from advertising. The famous actors of the Web, Google, Twitter and Facebook follow this line. Others reinvent the French "kiosque" and collect the applications developed by engineers or works realized by artists, which they sell to their large customer basis, giving back a share to the authors. iTunes is an example of the latter model.

OTT services are enabled by the delayering of the industry. Internet protocol (IP) has separated carriage from content and allowed "over-the-top" content and application providers to deal directly with end users over networks whose owners and operators are excluded from these transactions. The move to LTE's all-IP architecture will create a more open environment for these OTT providers and third-party services.

It is not only telecommunications that is affected. Internet television over broadband fixed and mobile networks is destabilizing the existing broadcasting industries. The gold mine of advertising is now also irrigating Internet players, and the trend is in their favor, the revenues growing with the number of people connected to the Internet. Newspapers have also experienced the competition with Internet news and its dramatic consequences.

In 2014, video will comprise 90% of Internet traffic and 64% of mobile data. Video and other OTT services are creating challenges for service providers. They must surpass subscriber quality expectations to acquire and retain market share by minimizing network-related artifacts and maximizing the subscriber quality of experience (QoE) regardless of which last mile technology is used or the target device.

### 3.1.2. *OTT applications*

Focusing now on the Internet, an OTT application is any app or service that provides a product over the Internet and bypasses traditional distribution. Services that come OTT are most typically related to media and communication and are generally, if not always, lower in cost than the traditional method of delivery.

The creation of OTT applications has led to a wide-ranging conflict between companies that offer similar or overlapping services. The traditional Internet Service Providers (ISPs) and telcos have had to anticipate challenges related to third-party firms that offer OTT applications. Think, for example, of the conflict between a company like Netflix and a cable company. Consumers still pay the cable company for access to the Internet, but they might get rid of their cable package in favor of streaming video over the Internet, which is cheaper. While the cable company wants to offer fast downloads, there is an inherent conflict of interest in not supporting a competitor, like Netflix, that bypasses cable's traditional distribution channel.

### 3.1.3. *OTT over LTE*

4G mobile networks claim that they provide to their customers the same performance as ADSL or Very high bitrate Digital Subscriber Line (VDSL), and will soon compare with Internet over coaxial cable and even with optical fiber to the home when LTE Advanced gets its

100 MHz. 4G clearly has the capability to offer High Definition Digital Television (HDTV) (US:720p and Europe:1080i). LTE can also offer 1080p, i.e. a better image definition, especially with H265. The only limit will be the capability of television sets to cater for ultra high definition. With the tsunami created by the switch from analog television to digital television, most homes have acquired flat panels displaying video on screens having a definition of 720p or 1080p. Normally, the renewal delay for a television set is 10 years or more; however, the price of the ultra High Definition (HD) (8 megapixels) devices is decreasing rapidly. New generation of set-top box is integrating 4 K and H265 for Q4 2014.

*OTT business*

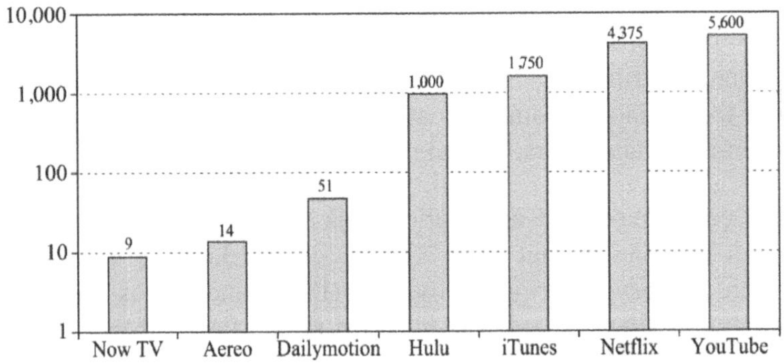


**Figure 3.1.** *2013 turnover of major OTT players in M$ (logarithmic scale)*

### 3.1.4. *New services opened by the high-speed Internet generalization*

When telecom operators offered their subscribers ubiquitous broadband access and data flat-rates, they opened the door to the digital society. However, with the advent of tablets and smartphones, subscribers now realize the enormous range of new possibilities, leading to a huge demand for innovative services that leverage the full capabilities of these devices. As a result, new ways of distributing information and content are changing the established business models, and telecom operators are in the pole position to capitalize by leveraging their large subscriber base.

To do so, content must be delivered to subscribers over existing IP networks with the same high quality known from established channels, but with the flexibility only the Internet can offer.

## 3.2. Technical view of OTT services

OTT content refers to the delivery of video, audio and other media over the Internet without a multiple system operator being involved in the control or distribution of the content. The provider may be aware of the contents of the Internet Protocol packets but it is not responsible for, nor able to control, the viewing abilities, copyrights and/or other redistribution of the content. This is in contrast to purchase or rental of video or audio content from an Internet service provider (ISP), such as pay television video on demand (VOD) or an IPTV video service, like AT&T U-Verse. OTT in particular refers to content that arrives from a third party, such as NowTV, Netflix, WhereverTV, NetD, Hulu, Crackle, WWE Network, RPI TV or myTV, and is delivered to an end user device, leaving the ISP responsible only for transporting IP packets.

An online video distributor (OVD) is defined as "any entity that offers video content by means of the Internet or other Internet Protocol (IP)-based transmission path provided by a person or entity other than the OVD".

Consumers can access OTT content through Internet-connected devices such as desktop and laptop computers, tablets, smartphones including iPhones and Android phones, set-top boxes such as the Roku and Google TV smart TVs, and gaming consoles such as the Wii, PlayStation 3 and Xbox 360. Consumers can access apps in most app stores.

### 3.2.1. *OTT technology*

OTT technology must allow multimedia (navigation into data, audio and video) diffusion without intervention of the ISP and mainly without a lot of integration between the service provider (app) and the device that receive the service (e.g. smartphone, smart TV and Set Top Box (STB)).

This is possible due to:

– common used software development kit (SDK);

– Digital Rights Management (DRM);

– streaming protocol;

– back office solutions.

3.2.1.1. *Commonly used SDK*

The objective of a software development kit (SDK) is to provide application developers with the tools to make their app. When the launch of an SKD is a success it creates a community of developers and large offer of apps that will create more demands on the devices (smartphones, TV, etc.) that support this SKD.

Today, the two main SKDs seen on the market are:

– iPhone OS (iOS) SKD for iPad (https://developer.apple.com/devcenter/ios/index.action);

– Android is an open source operating system (http://developer.android.com/sdk/index.html). Google allows the manufacturer integrating Android to modify it and bring their personalization to the operating system. It is based on WebKit, OpenGL, SQLite ouFreeType. The core of Android is based on Linux, modified to be adapted for mobile and using a Java virtual machine (Dalvik).

HTML5 (http://www.w3.org/TR/html5/) is used largely on the Web for the development of websites that include gaming or video (flash is decreasingly used in 2014). This very open architecture enables the developer to reuse their code on any device that is compatible with HTML5 (STB, TV, tablet and phones). The standards have evolved to include the management of the video (tag <video> <audio>), the DRM (e.g. encrypted media extensions (EMEs) for DRM interoperability (i.e., talking to PlayReady or WideVine) as well as ClearKey, which is a lighter-weight encryption protocol), communication between devices (DIAL) and the live streaming (browser extensions MediaSource Application Programing Interface (API) (MS)).

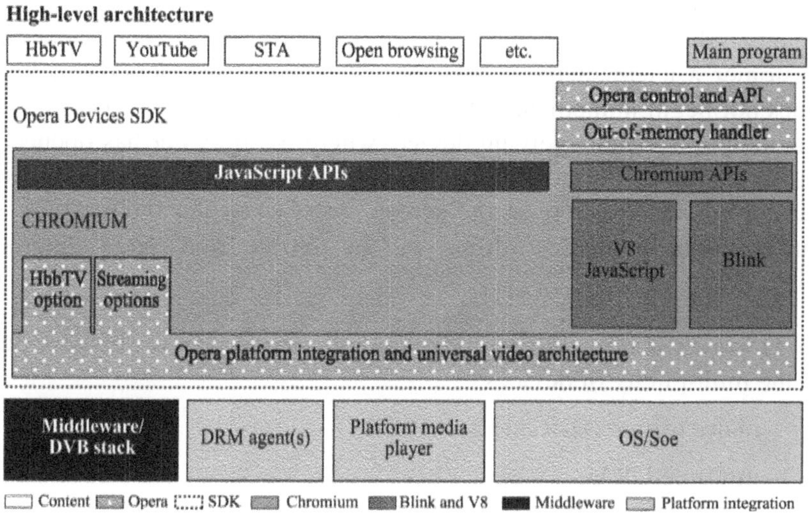


**Figure 3.2.** *Architecture HTML5 by Opera (source: www.operasoftware.com)*

### 3.2.1.2. *DRM management*

Some issues are not easy to solve, such as digital rights management with copyright, download, retransmission, etc. The best example is Web television, which has been a reality for several years. YouTube and Dalymotion are key players in Web TV. In France, TDF Media Service also has a multimedia platform providing multimedia content to any kind of terminal. They started the service without DRM for free content. Today the standard of HTML5 and the player of the iOS and Android operating system (OS) include DRM and let the player have the possibility of developing their paid TV offer (https://developers.google.com/youtube/branding_guidelines) with protection of the content. One example is the YouTube new Leanback service that offers security of the service with Playready DRM or

Widevine. The integration of the DRM is not as difficult as it used to be.

### 3.2.1.3. *Streaming protocol*

Since the OTT operator has not achieved the QoS of the Internet channel, the technology makes use of buffering and caching: when the customer asks for a movie, the provider begins the download. Play does not begin immediately; the beginning of the movie is recorded in the memory of the receiving device, where the provider has opened a buffering area without noticing. That buffer will allow a fluid flow of the movie presentation on the screen, even when the Internet link is suffering from delays, drops in the bitrate and other Internet malfunctions.

However, the real difficulty in finding a good way to stream video is the multiplicity of devices:

– many format ABR and non–ABR;

– multiple DRM;

– multiple formats of subtitles;

– format that requires increased storage capability (buffering and records);

– transcodage that increasingly requires Central Processing Unit (CPU) (e.g. HEVC encoding today requires twice CPU than HD H264).

In 2012, the formats largely used in the market were Apple HTTP live streaming (HLS), Widevine Media Format (WVM), Adobe Http Dynamic Streaming (HDS) and Microsoft Http Smooth Streaming (HSS). In 2014, MPEG DASH seems to be the unified protocol. Whereas we see the first deployment of HEVC and demonstration of 4 K, the following streaming protocols are used:

– main: H.264-Advanced Audio Coding (AAC) using Dash TS (broadcast) and DASH fMP4 (ott) HEVC/AAC in DASH fMP4;

– legacy (without the support of HEVC): HLS/HSS/MP4 HTTP.

DASH is a new protocol that will soon lead OTT services:

– it is integrated into thousands of commercial OTT solutions or open-source players;

– big groups are participating in this standard, such as Microsoft, YouTube and Netflix;

– Widevine has abandoned its packaging proprietary WVM for DASH;

– the Dash Industry Forum is favorably aggregating the industry player;

– benefits of technology breakthrough of HEVC and 4 K.

However, it is not yet largely deployed:

– more in an evaluation step;

– loss of deployment used ABR and there is a cost to migrate to DASH;

– HEVC is not yet deployed and looks to be the good reason to migrate to DASH.

The player installed in the smartphone is well-positioned in the market of the HEVC/Dash player because of the support of HEVC in the chipset Qualcomm Snapdragon 800.

#### 3.2.1.4. *OTT back office*

The components of a back office are:

– VOD servers;

– liver streaming servers;

– encoders;

– business management system (BMS);

– DRM protection;

– Cas protection;

– profiling;

– multiscreen manager;

– resource manager;

– delivery network;

– connected devices.

The encoders are led by the evolution of the video format and streaming protocol that evolves constantly.

Encoder and packager are doing a repacking of the live video stream (TS) to ABR or all formats that are supported on the fields.

The big family of encoders are:

(VOD) ABD=> ABR;

(VOD) MP4=> ABR;

(VOD) Mezzanine=>ABR;

(Live) ABR=> ABR;

(Live) RTMP=>ABR;

(Live) TS=>ABR.

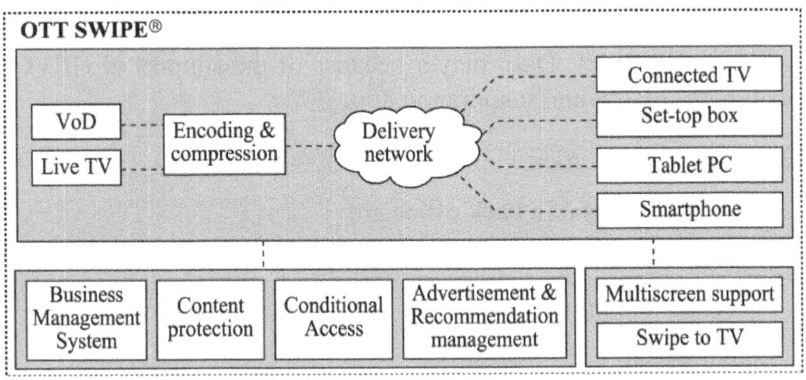


**Figure 3.3.** *OTT SWIPE® – over-the-top TV for telecom operators*

Siemens Over-The-Top SWIPE® (OTT SWIPE®) utilizes Web-based technologies in order to create a high-quality viewing experience by streaming videos OTT through already existing networks. Subscribers can watch these videos on a variety of connected devices such as connected TVs, STBs, tablets, smartphones and personal computers. It is the natural extension of an IPTV offering because OTT SWIPE® the operator leverages IPTV content on networks that are outside of its control. Hence, an operator directly accesses its competitors' subscribers. OTT SWIPE® runs via apps and browsers on the devices that subscribers already own, and it is very intuitive to use. At the touch of their fingertips, subscribers can conveniently swipe content from one device to another – it puts the magic back into the TV experience. OTT SWIPE® features a BMS that optimizes the end-to-end lifecycle of available content, and it enables all types of business models, such as pay-per-view or subscriptions. Briefly, it makes it possible to leverage a large existing subscriber base with high quality and high-value additional video services in order to create a new, sustainable revenue stream.

### 3.2.2. *Testing OTT performances*

Testing the delivery of OTT for performance, availability, security and scalability (PASS) allows engineers to answer the following questions:

– Can the server and the network support the anticipated number of users?

– What is the end-to-end QoE for any of the thousands of users?

– How does the service perform under adverse network conditions, such as latency and loss?

– How does security affect video performance and QoE?

– Does the HTTP adaptive streaming solution deliver the appropriate bitrate encoded video for the destination device and upshift or downshift as requirements change?

– Does packet loss happen across all network failure scenarios to reduce the cumulative visual effect on all subscribers?

## 3.3. OTT services challenging TV telecommunication services

OTT companies can use the Web for offering telecommunications services, such as SMS, MMS, Voice over the Internet Protocol (VoIP), text, images and group calls, directly challenging the services provided by the network operators, which have higher prices, since they have invested a huge amount of capital in building the infrastructure. Network operators claim they are "pirates". There is a correlation between the blossoming of telecom OTT and the call marketing and call advertising business.

OTT messaging refers to a similar idea, where a third party provides instant messaging services as an alternative to text messaging services provided by a mobile network operator.

The most well-known services are:

– *Skype* is the most famous OTT service for telecommunications. Initially a Dutch start-up, it has since been bought by Microsoft. The services are VoIP and visioconferencing. Millions of users worldwide, mostly students and families, have adopted Skype.

Internet telephony, or VoIP, is the first OTT service with major implications for the business models of both fixed and mobile network operators.

**Figure 3.4.** *Skype*

More recently, text messages (SMS) have also been delivered OTT affecting the revenues of fixed and mobile operators.

– *WhatsApp Messenger* offers messaging services such as SMS via the Web. Customers of the service can be connected by Wi-Fi or mobile systems. Provided the sending party owns the same application, messages can be received on iPhone, Android, Blackberry, Windows Phone, Symbian, etc. Billions of messages are sent by Whatsapp every day.

**Figure 3.5.** *WhatsApp*

*Viber* has about the same offer as Whatsapp. It provides audio but not video. Viber claims over 100 million users worldwide.

**Figure 3.6.** *Viber*

In Europe, operators have recently been challenging the right of OTT companies to offer services, without practical results to date. US operators prefer to negotiate with them in order to recover some revenue for the traffic that is traveling on their links. Technically, the network operator has the possibility of squeezing the packets from these services, even if it is very difficult – this does not give a blackmail position but at least a starting point for discussion.

### 3.3.1. *Instant messaging business*

Instant messaging apps are a potent force in the mobile ecosystem. OTT messaging services and apps work over an Internet connection and not over a wireless network like standard SMS texts do.

OTT messaging has grown exponentially since 2011, when Informa and Portio Research both pegged its volume at about 4 billion daily messages. By year-end 2013, daily OTT message traffic was expected to be more than double the amount of daily SMS texts.

Clearly, increasing smartphone penetration is the catalyst behind this surge. The proliferation of mobile messaging services has also played a role.

The largest players are only getting larger: Apple's proprietary iMessage service now delivers over 2 billion daily messages, while WhatsApp sports over 200 million active users. Smaller mobile messaging services geared toward younger demographics, like MessageMe and Snapchat, have also enjoyed popularity.

The next step for the OTT players is to effectively monetize their product, be it by embracing advertising, or through paid premium services and subscriptions.

Wireless carriers, on the other hand, must combat the OTT threat head-on, since it may siphon revenue from their SMS text business at a greater rate than originally projected. Some carriers have introduced their own OTT apps. Others may gravitate toward partnerships with OTT services, or lean more heavily on charging for data instead of SMS and voice packages.

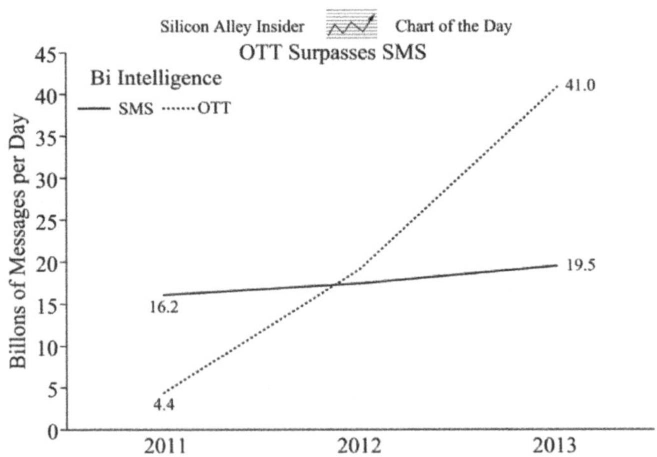


**Figure 3.7.** *Compared traffic SMS versus OTT*

### 3.3.2. *Television and video OTT services*

OTT television refers to video, television and other services provided over the Internet rather than via a service provider's own dedicated, managed IPTV network.

OTT is delivered directly from provider to viewer using an open Internet/broadband connection, independently of the viewer's ISP, without the need for carriage negotiations and without any infrastructure investment on the part of the provider. It is a "best effort", unmanaged method of content delivery via the Internet that suits providers who are primarily broadcasters rather than ISPs.

OTT opens up a completely new way for operators to generate revenue from VOD, catch-up TV and interactive applications. The clearest opportunity is for traditional TV distributors. With established brands, strong relationships and consumers' trust, they are perfectly placed to deliver a seamless service that integrates online and traditional TV through a single device managed by a single operator.

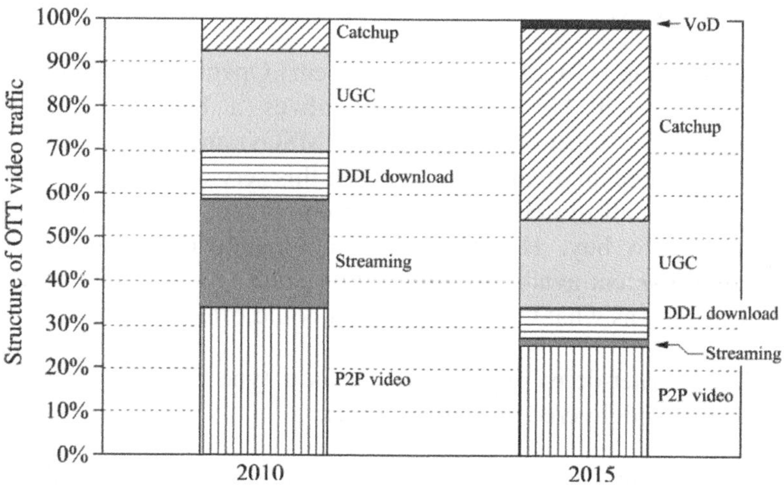


**Figure 3.8.** *Video on Internet traffic (source: http://meta-media.fr/2012/07/12/la-video-represente-deja-90-du-trafic-internet-en-france.html)*

Because of its openness and perceived ability to bypass the networks owned and operated by payTV providers, OTT is sometimes described as a potential "killer" for the payTV industry. We believe that this is a naive viewpoint. The rights to quality video content are expensive, and such content simply will not be created unless revenue can be generated from it. At present, the content industry is supported by subscription TV – and its delivery medium is the set-top box or residential gateway.

The software in the STB allows the content of OTT. The appeal of OTT is that it augments the traditional viewing experience by allowing the user to access Internet-based content without having to leave their TV set.

OTT will continue to play an important role in allowing consumers to access their paid-for services in the way they want, adding value to payTV subscriptions. Ultimately, consumers will enjoy content "any time, any place" through a range of devices including mobile phones, iPads, games consoles or connected TVs, all connected to a central content gateway.

Over-the-top virtual Multiple System Operator (MSO) is an Internet-delivered TV service that utilizes a broadband Internet connection to compete with cable TV (MSOs) and certain other TV providers. There are many advantages to this service. There is no need for a service call to deliver and set-up equipment, and potentially no set-top box to buy. However, there are bandwidth costs and the amount of content available using *OTT virtual MSOs* would typically not be as high as what the cable provider could offer.

This Internet-delivered TV service is a type of subscription TV service that is delivered over the Internet.

Over-the-top Home Entertainment Media are electronic devices that manufacturers are providing. They are DVD players, video game consoles and TVs with built-in wireless connectivity. These devices piggy back on an existing wireless network, pull content from the Internet and deliver it to the TV set. Typically, these devices need no additional wires, hardware or advanced knowledge on how to operate. Content suited for TV can be delivered via the Internet.

### 3.3.3. *Apple TV (source: Wikipedia)*

Apple TV is a network device that allows consumers to use an HDTV set to view photos, play music and watch video that originates from an Internet media service or a local network.

Apple TV is a digital media player developed and sold by Apple Inc. It is a small network appliance and entertainment device designed to play digital content from the iTunes Store, Netflix, Hulu Plus, YouTube and Vevo, along with the TV Everywhere portals of several cable and broadcast networks, and the video subscription portals of three of the four major North American sports leagues: MLB.tv, NBA League Pass and NHL GameCenter. The device also plays content from any Mac OS X or Windows computer running iTunes on an enhanced-definition or high-definition widescreen television.

Apple offered a preview of the device in September 2006, and began shipping it the following March. It initially shipped with a 40 GB hard disk; a 160 GB version was introduced 2 months later, and the earlier model was ultimately discontinued.

In September 2010, Apple announced the second-generation version of the Apple TV. About one-quarter of the size and one-third of the price of the original Apple TV, the new device could stream rented content from iTunes and video from computers or iOS devices via AirPlay. This version no longer had a hard drive; however, it does have an undocumented internal 8 GB flash storage, speculated to be used for smoother playback of streamed media. All content is drawn from online or locally connected sources.

The third generation of the device was introduced at an Apple event on 7 March 2012, with new features such as higher resolution (1,080 p) and a new user interface.

On 28 January 2013, Apple released a third generation "Rev A" that included component changes.

Notable competitors include WD TV, Roku, Amazon Fire TV, Google TV, and Chromecast, consoles and media hubs such as the PlayStation 3/4, Xbox 360/One and Nintendo Wii/WiiU and late model TiVo DVR systems, as well as internal smart TV systems and Blu-ray players from companies such as Vizio, Sharp, Sony, Samsung, LG and others.

Users can access the iTunes Store directly through Apple TV to rent movies and TV shows and stream audio and video podcasts. While the first generation of the Apple TV could download content, the second generation lacks a hard drive, and thus cannot store purchased content. Users who wish to purchase content on the Apple TV may do so, but cannot download directly to the Apple TV. Content must be streamed live or downloaded via iTunes onto a device with storage capability (personal computer, iPhone, iPad, etc.). Since 2008, podcasts have been served on the Apple TV as other kinds of video, as opposed to in Really Simple Syndication (RSS) and similar feeds. Until mid-March 2009, the Apple TV was the only way to purchase HD iTunes content.

Apple TV can display photos from Flickr and iCloud in a slide show, with automatic cross-dissolve transitions, and optionally with the Ken Burns effect.

Netflix streaming and Hulu Plus integration are available, as well as YouTube and Vimeo video via included apps.

Quite a few television programs are available, e.g. NBA TV, Disney Channel, ESPN, HBO, Weather Channel, Bloomberg, History and Lifetime.

Parental controls allow consumers to limit access to Internet media service content, via the restrictions setting; individual services can be turned off (e.g. to reduce clutter), and their icons can be rearranged via the tap-and-hold technique in iOS. Internet media is split into four categories: "Internet Photos", "YouTube", "Podcasts" and "Purchase and Rental". Each of the categories is configured by a parental control of "Show", "Hide" or "Ask" to prompt for a four-digit preset code. In addition, movies, TV shows, music and podcasts can be restricted by rating.

Apple TV can sync or stream photos, music and videos from a computer running iTunes.

A user can connect a computer to a local network to maintain a central home media library of ripped CD, DVD or HD content, provide direct connectivity to photo organization software, such as iPhoto, limit home video access to a local network only, play Internet radio or preload content on Apple TV to be used later as a non-networked video player. For users who wish to connect the Apple TV to a computer, synchronization and streaming modes are supported.

Apple TV in synchronization mode works in a way similar to the iPod. It is paired with an iTunes library on a single computer and can synchronize with that library, copying all or selected contents to its own hard drive. Apple TV need not remain connected to the network after syncing. Photos can sync from iPhoto, Aperture or from a hard disk folder on a Mac, or Adobe Photoshop Album, Photoshop Elements, or from a hard disk folder in Windows.

Apple TV can also function as a peer-to-peer digital media player, streaming content from iTunes libraries and playing the content over the network.

### 3.3.4. *Netflix, the 2014 OTT champion*

You can think – from telcos or cable/satellite companies say – that OTT aims at replacing your regular TV provider. Examples include Hulu or Netflix.

Netflix is an American company that offers movies on the Internet as a continuous flow. It operates in Northern and Southern America, in the Carribean and in some European countries, such as UK, Ireland, Netherlands and the Scandinavian countries.

Netflix also provides rented DVDs by mail order. The provision of movies in continuous flow started in 2010. It can be accessed with computers having Windows or Mac OS, and with other compatible devices, e.g. Xbox, PS3 and Wii. Also compatible are connected television sets from Samsung and LG.

In 2009, the Netflix offering counted more than 100,000 movie titles. The customer base was already over 11 million; it reaches now more than 40 million. Over 30% American homes have a Netflix subscription (8$/month). Netflix is responsible for more than 30% of Internet traffic in Northern America.

Since 2013, Netflix proposes a collaborative filtering service. The collaborative filtering algorithm analyses the user's habits. This service mobilizes more than 900 engineers and is used to focus the offers on the real preferences of each customer.

Recognizing the growing success of Netflix, Disney and other media producers are now producing some television movies especially for Netflix.

### 3.3.5. *"OTT services" provided by the network operators*

OTT is a general term for a service that you utilize over a network that is not offered by that network operator. It is often referred to as "over-the-top" because these services ride on top of the service you already get and do not require any business or technology affiliations with your network operator.

However, Sprint is considered an "over-the-top long distance service" as it primarily offers long distance over other phone company's phone lines. Often, there are similarities between the service your network operator offers and the service that the OTT provider offers.

Nothing prevents the network operator designing and marketing its own OTT services to take advantage of the proliferation of Internet television and Internet-connected TVs.

In France, the mobile operator Bouygues Telecom (ByT) takes advantage of its 4G LTE network to offer OTT services (in conjunction with Cognacq-Image, which has a wide experience of video and work flows). ByT has been granted a competitive advantage by the French regulatory authority *Autorité de regulation des communications électroniques et de la poste* (ARCEP): it is the only operator that has been allowed to roll out LTE in the 1,800 MHz band. That frequency band is common with the US allocation, and iPhone accesses this band in 4G while it does not connect to 800 MHz and 2.6 GHz eNodeB of ByT competitors.

This OTT offer will have more than 60 movies available. It is intended to grow up to the actual offering of 200 different programs that ByT provides on the ADSL in IPTV.

### 3.3.6. *The carrier: neutral or responsible?*

The ISP is supposed to be unable to control the content, acting merely as the "dumb pipe" that transports the bits. This altruistic vision, in which Net Neutrality rules and ISPs turn a blind eye to content being delivered on its pipe, is easily shattered. Not only are ISPs being asked to monitor for copyright infringements, but they are requesting the ability to use deep-packet detection to do so.

The latest solution – decided among the ISPs and representatives from the Motion Picture Association of America (MPAA) and the Recording Industry Association of America (RIAA), after public backlash to congressional support for the Stop Online Privacy Act (SOPA) – requires the ISPs to monitor for copyright infringements.

Initially intended to launch in mid-2011, then delayed to the end of 2011, then delayed to mid-2012, then again to late 2012, the new policy is called the Copyright Alert System (CAS). CAS was set to go into effect on 28 November 2012, the latest delayed launch date, but the Center for Copyright Information (CCI) pushed back the start date again, citing a hurricane as the reason.

"Due to unexpected factors largely stemming from Hurricane Sandy," wrote Jill Lesser, executive director of CCI, "which have seriously affected our final testing schedules, CCI anticipates that the participating ISPs will begin sending alerts under the Copyright Alert System in the early part of 2013, rather than by the end of the year."

The CAS will result in verbal and written warnings to ISP customers, purportedly escalating from nice to nasty, and the sixth infraction will incur something. What it will incur is up to the individual ISP, and there may be little more than bandwidth throttling from ISPs that do not have "skin in the game" when it comes to content ownership. For those major ISPs, such as Time Warner Cable, Inc. (TWC) that also owns premium content, the infractions could lead to an ISP user's account being terminated.

The "six strike" policy is based on a similar "three strike" model that went into effect in New Zealand in which notices move from detection to warning to enforcement.

Interestingly, ISPs reported lower traffic when the new law went into effect in late 2011, with Orcon – a local New Zealand ISP – noting that its international traffic had dropped by around 10%.

Yet, the first case brought by the RIANZ, the New Zealand equivalent of RIAA, was dismissed in court in October 2012, after the accusation against the ISP user proved to be unsubstantiated. Apparently a student, who shared a flat with other students, had signed up for an ISP account, and someone in the flat – or accessing the unsecured Wi-Fi network – used file-sharing software to download copyrighted content. It turns out that the student did not even know what file-sharing software was, as Tech Liberty NZ noted in its request for dismissal of the student's case.

"The law is meant to act as a deterrent to infringing copyright," Tech Liberty notes, "but the way it is written it is actually an incentive. 'Just use a connection that doesn't have your name on the account and they'll be the one who is penalised!' The only deterrent is to becom[e] an internet account holder."

## 3.4. OTT services other than television

OTT offerings include thousands of applications, which do not monopolize important bitrates on the Internet.

### 3.4.1. *Dedicated services*

With LTE, the bandwidth will increase and provide the opportunity to use more comfortable services in the cloud: mobile cloud computing (MCC) with storage in the cloud, backup, deported user interface.

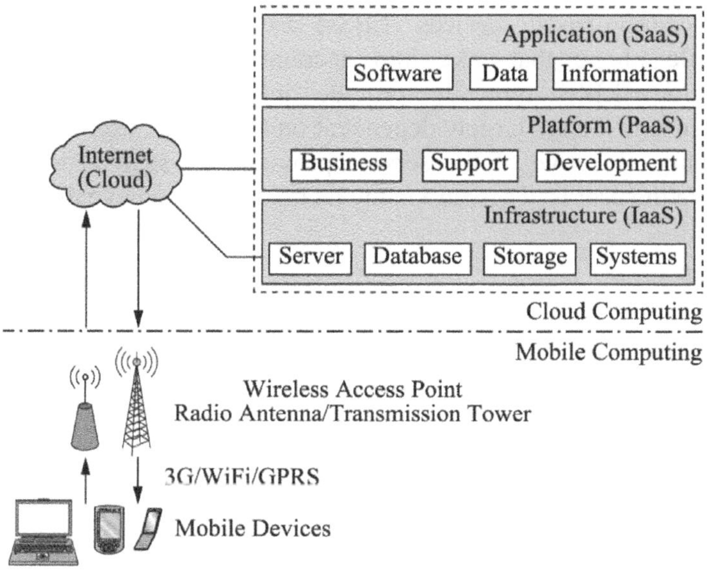


**Figure 3.9.** *Mobile cloud computing*

MCC in its simplest sense refers to an infrastructure where both the data storage and the data processing occur outside the mobile device. Mobile cloud applications move the computing power and data storage away from mobile phones and into the cloud, bringing applications and mobile computing not just to smartphone users but to a much broader range of mobile subscribers.

Assign the appropriate pricing mechanism to maximize the revenue of MCC systems and provide incentives for mobile users, considering different cases, e.g. resource may be stable but applications may transmit to other places.

The Internet of things will in the same trend be easier, adding mobility to the ecosystem.

### 3.4.2. *LBS: positioning and GPS-driven applications*

Location-based services (LBSs) are services offered through a mobile phone that take into account the device's geographical location. LBSs typically provide information or entertainment. Because LBSs are largely dependent on the mobile user's location, the primary objective of the service provider's system is to determine where the user is. There are many techniques to achieve this.

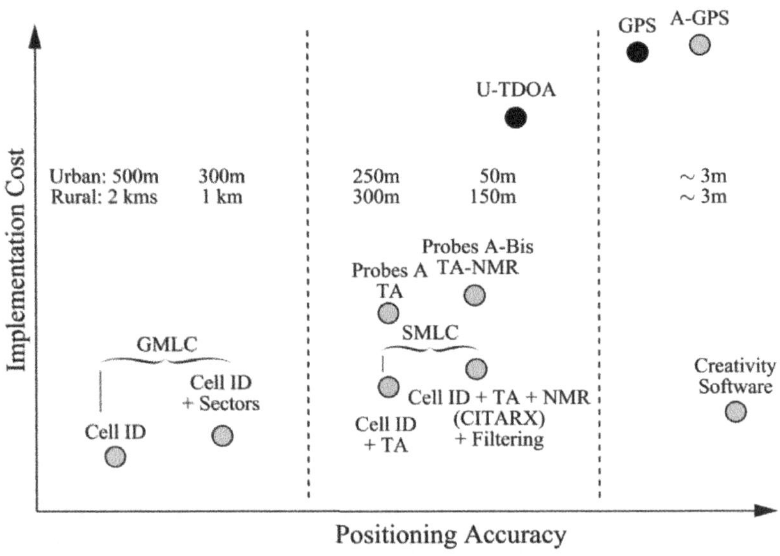


**Figure 3.10.** *LBS*

The location is provided either by the mobile network, which knows the mobile position from the recorded parameters that allow paging an idle mobile or those that are more accurate obtained by triangulation. Only Global Positioning Satellite (GPS)-based information is accurate enough to support LBS.

The LTE system optionally includes location servers, serving mobile location center (SMLC) and gateway mobile location center (GMLC), interfacing the LBS server.

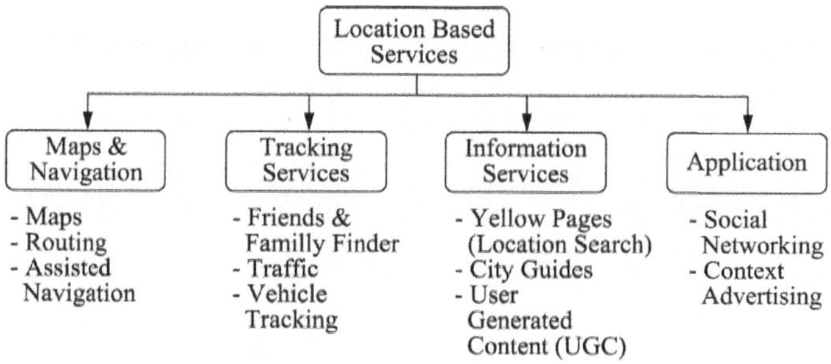


**Figure 3.11.** *Location-based services*

Some of the most common LBS applications include local news, directions, points of interest, directory assistance, fleet management, emergency, asset tracking, location-sensitive building and local advertisement. In fact, location services cover a wide range of services, orientated at individual users or devices and vehicles.

These can include:

– "Find Me" services e.g. nearest restaurant, coffee bar, ATM etc.;

– location-based charging, for example, "Home Zone" or "Office Zone";

– retail offers – dependant on user proximity;

– social networking services;

– visitor local information services;

– child find/locate services;

– route planning/navigation;

– "Where am I" type information services;

– location-based games;

– vehicle and asset tracking;

– support for emergency services and law enforcement agencies.

Such applications may be supported within a mobile operator's network, or externally via third parties. In both cases, the application that depends on information obtained and processed by network resources is referred to as a Location Services (LCS) client.

An LBS, for example, can point a user to the nearest restaurant. In another example, an LBS can send an SMS message advertising a sale at a nearby shopping mall.

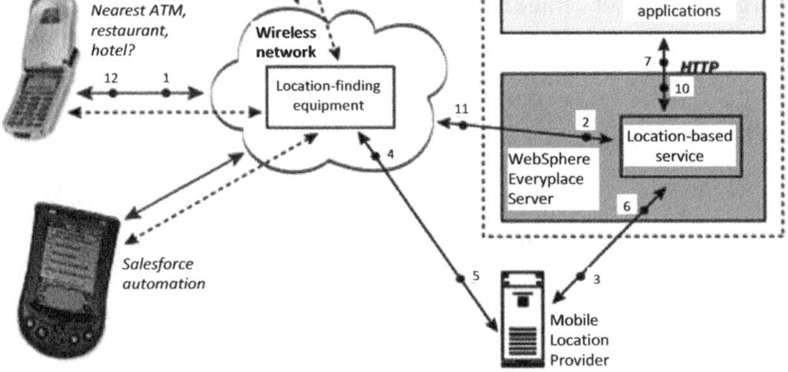


**Figure 3.12.** *Location services architecture*

These services can be classified into two types: push and pull. In a push type of service, the user receives information from the service provider without requesting it at that instant, although the user may have originally subscribed to the service at an earlier time. The LBS advertisement mentioned earlier is an example of a push service. In a pull type, the user has to actively request information. The restaurant query example belongs to this type.

Software development platforms, particularly those used for creating mobile applications, such as J2ME and Android, have specialized APIs that support LBS.

### 3.5. Open applications versus verticalization

The business model for OTT is not optimal when only based on using the operator's infrastructure for applications the users are ready to pay for. As shown earlier, the operator has the possibility to develop and market the same kind of service. The way to make a maximum profit seems verticalization with non-standard applications and devices.

#### 3.5.1. *The Apple model*

Apple is the paramount example of verticalization. This strategy has been enforced by Steve Jobs along successive steps:

– introduction of the iPod for music listening: the iPod does not accept music and sounds compressed with the celebrated MP3 compression algorithm, which is the standard MPEG-1 Audio layer 3. In order to keep its customers, Apple has created a library of music works encoded with the standard AAC (MPEG-2-Part 7 and MPEG-4-Part 2) with a particular version called iTunes Plus;

– introduction of the iPad and iPhone, based on the same OS in order to share applications: iPad, iPhone, iPod Touch and iMac share the same platform for applications uploading, called App Store. This software is integrated into the iTunes Store service. It allows us to upload applications on all Apple products.

The (iTunes) App Store has been a phenomenal success from its introduction in 2008, followed by the (Mac) App Store in 2011. Today, there are some one million applications in the App Store that generated in 2013 far more than 50 billion downloads, a turnover of about 18 billion dollars and a 4.2 billion margin.

The other side of the Apple system is the SDK for developers. Developers must submit their work to the checking process of Apple. Apple, therefore, controls completely the App Store content. The developer decides the price of its application and receives 70% of the revenue generated by this application.

# Conclusion

## C.1. Introduction

The world's first publicly available LTE service was launched by TeliaSonera in Stockholm and Oslo on 14 December 2009. In fact, this opening was more a demonstration of the manufacturer Ericsson showing that it was able to engineer a viable system implementation. Just remember that in 2009 LTE had no solution for voice telephony call except OTT VoIP.

LTE is the natural upgrade path for carriers with both GSM/UMTS networks and CDMA 2000 networks such as Verizon Wireless. Verizon launched the first large-scale LTE network in North America in 2010. The other big CDMA 2000 player KDDI announced that it will migrate to LTE. Airtel launched the LTE service in India in April 2012. In Europe, effective deployments of LTE began slowly after the decision about the 800 MHz frequency band pulled from the broadcasting world by the decision of the European Commission on 6 May 2010, but transcribed in European law around June 2011. European operators are probably waiting for the opening of the 700 MHz frequency band with a stable plan which is still not yet available. Fortunately, the final decision on this frequency band will intervene in 2015 at the WRC 15 (Geneva).

China is not pushing the deployment of LTE hard. Unlike most other large economies, China opens networks based on TDD and at 2.4 GHz, which means for dense urban environment. This behavior could be considered surprising since the most advanced manufacturer for LTE seems to be HuaWei from ShenZhen, which is supplying many LTE operators around the world, except in the USA where this company has been forbidden to make offers.

At the end of 2012, 104 countries had announced that they were adopting LTE. Unsurprisingly, it is the only real concretization of IMT/IMT Advanced. Of these 104, 51 are announcing commercial networks. This point is followed by the Global Mobile Suppliers Association (GSA), the association of mobile manufacturers.

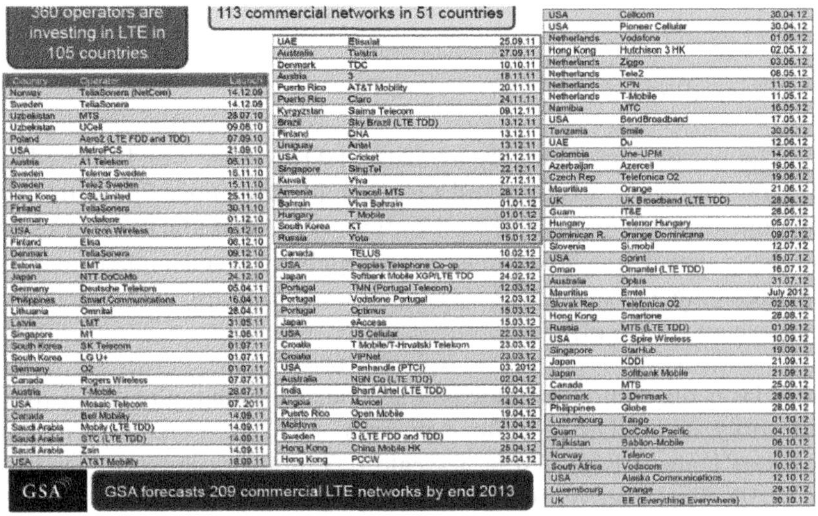


**Figure C.1.** *LTE forecast by the end of 2013*

LTE is therefore anticipated to become the first truly global mobile phone standard, although the different LTE frequency bands used in different countries will mean that only multi-band phones will be able to use LTE in all countries where it is supported. The issue of multiband operation has been addressed here from a technical point of view.

Even in Europe, LTE will be deployed in quite a few frequency bands.

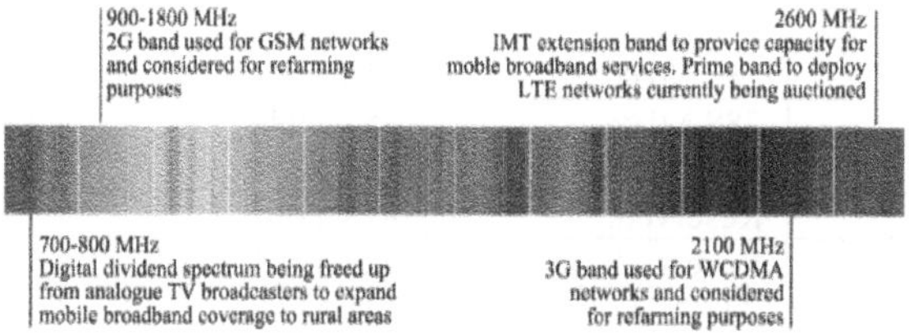


**Figure C.2.** *Band of frequencies*

This presents a difficult challenge. The technical answer might be software-defined radio devices taking advantage of the new progress of the micro/nanoelectronics offering still more processing power and hopefully a lower power consumption. Component manufacturers thus have a lot of work before them.

| E-UTRA Operating Brand | Uplink (UL) operating band BS receive UE transmit | Downlink (DL) operating band BS transmit UE receive | Duplex Mode |
|---|---|---|---|
| | $F_{UL\ low}$–$F_{UL\ high}$ | $F_{DL\ low}$–$F_{DL\ high}$ | |
| 1 | 1902 MHz–1980 MHz | 2110 MHz–2170 MHz | FDD |
| 2 | 1850 MHz–1910 MHz | 1930 MHz–1990 MHz | FDD |
| 3 | 1710 MHz–1785 MHz | 1805 MHz–1880 MHz | FDD |
| 4 | 1710 MHz–1755 MHz | 2110 MHz–2155 MHz | FDD |
| 5 | 824 MHz–849 MHz | 869 MHz–894 MHz | FDD |
| 6[1] | 830 MHz–840 MHz | 875 MHz–885 MHz | FDD |
| 7 | 2500 MHz–2570 MHz | 2620 MHz–2690 MHz | FDD |
| 8 | 880 MHz–915 MHz | 925 MHz–960 MHz | FDD |
| 9 | 1749.9 MHz–1784.9 MHz | 1844.9 MHz–1879.9 | FDD |
| 10 | 1710 MHz–1770 MHz | 2110 MHz–2170 MHz | FDD |
| 11 | 1427.9 MHz–1447.9 MHz | 1475.9 MHz–1495.9 MHz | FDD |
| 12 | 699 MHz–716 MHz | 729 MHz–746 MHz | FDD |

| | | | |
|---|---|---|---|
| 13 | 777 MHz–787 MHz | 746 MHz–756 MHz | FDD |
| 14 | 788 MHz–798 MHz | 758 MHz–768 MHz | FDD |
| 15 | Reserved | Reserved | FDD |
| 16 | Reserved | Reserved | FDD |
| 17 | 704 MHz–716 MHz | 734 MHz–746 MHz | FDD |
| 18 | 815 MHz–830 MHz | 860 MHz–875 MHz | FDD |
| 19 | 830 MHz–845 MHz | 875 MHz–890 MHz | FDD |
| 20 | 832 MHz–862 MHz | 791 MHz–821 MHz | FDD |
| 21 | 1447.9 MHz–1462.9 MHz | 1495.9 MHz–1510.9 MHz | FDD |
| 22 | 3410 MHz–3490 MHz | 3510 MHz–3590 MHz | FDD |
| 23 | 2000 MHz–2020 MHz | 2180 MHz–2200 MHz | FDD |
| 24 | 1626.5 MHz–1660.5 MHz | 1525 MHz–1559 MHz | FDD |
| 25 | 1850 MHz–1915 MHz | 1930 MHz–1995 MHz | FDD |
| 26 | 814 MHz–849 MHz | 859 MHz–894 MHz | FDD |
| 27 | 807 MHz–824 MHz | 852 MHz–869 MHz | FDD |
| 28 | 703 MHz–748 MHz | 758 MHz–803 MHz | FDD |
| … | | | |
| 33 | 1900 MHz–1920 MHz | 1900 MHz–1920 MHz | TDD |
| 34 | 2010 MHz–2025 MHz | 2010 MHz–2025 MHz | TDD |
| 35 | 1850 MHz–1910 MHz | 1850 MHz–1910 MHz | TDD |
| 36 | 1930 MHz–1990 MHz | 1930 MHz–1990 MHz | TDD |
| 37 | 1910 MHz–1930 MHz | 1910 MHz–1930 MHz | TDD |
| 38 | 2570 MHz–2620 MHz | 2570 MHz–2620 MHz | TDD |
| 39 | 1880 MHz–1920 MHz | 1880 MHz–1920 MHz | TDD |
| 40 | 2300 MHz–2400 MHz | 2300 MHz–2400 MHz | TDD |

| 41 | 2496 MHz–2690 MHz | 2496 MHz–2690 MHz | TDD |
|---|---|---|---|
| 42 | 3400 MHz–3600 MHz | 3400 MHz–3600 MHz | TDD |
| 43 | 3600 MHz–3800 MHz | 3600 MHz–3800 MHz | TDD |
| 44 | 703 MHz–803 MHz | 703 MHz–803 MHz | TDD |
| NOTE 1: Band 6 is not applicable | | | |

**Table C.1.** *UL/DL*

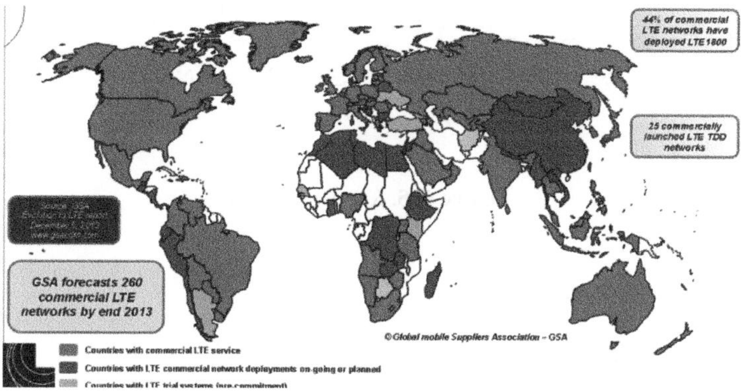


**Figure C.3.** *Adoption of LTE technology as of 5 December 2013.*

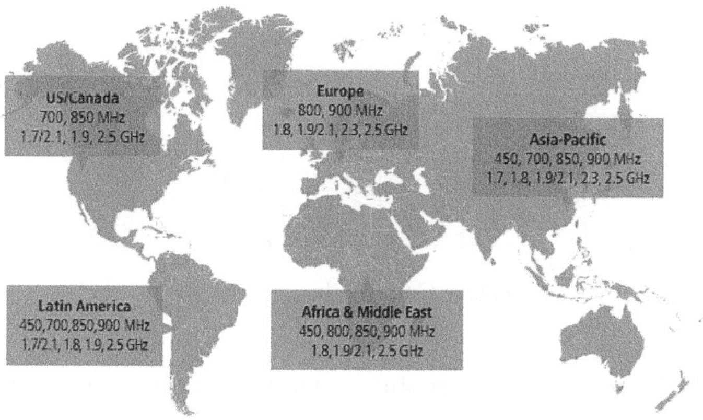


**Figure C.4.** *Forecast frequency bands around the world as of 2013*

The GSA provides a summary of all known projects to date.

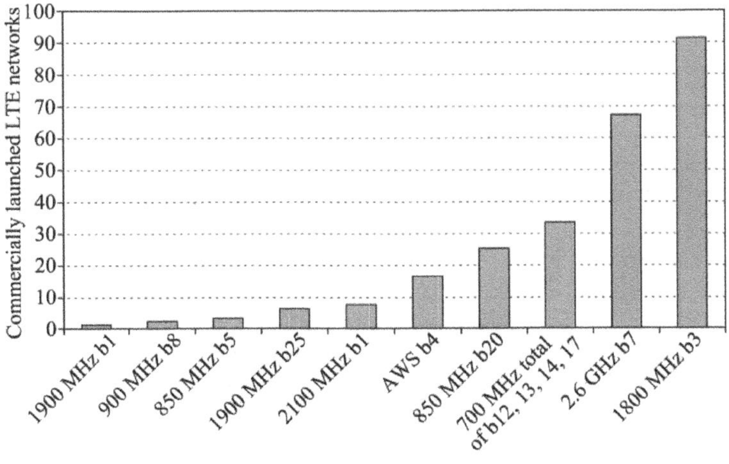


**Figure C.5.** *Known projects*

## C.2. Market forecast

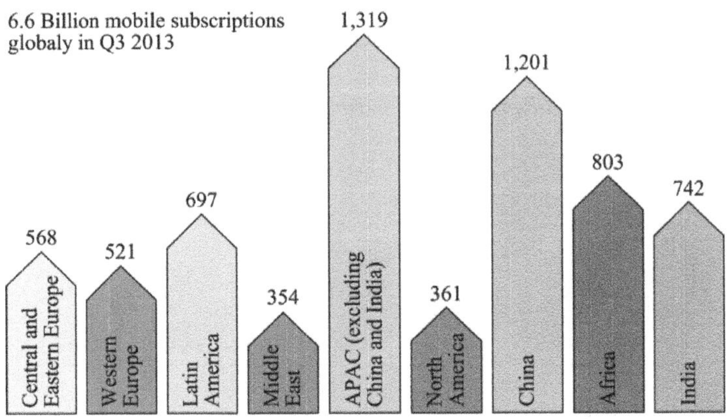


**Figure C.6.** *Mobile subscriptions*

Already in 2014, the mobile communications market could be considered saturated.

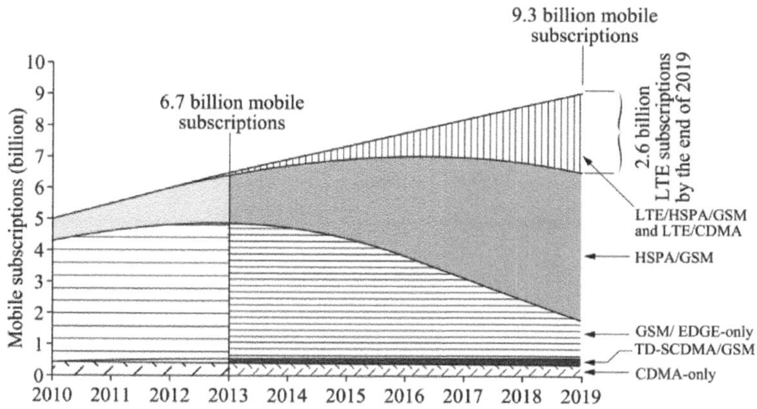


**Figure C.7.** *Mobile subscription forecast by the end of 2019*

Therefore, LTE will face in many countries a replacement market. This is not an issue: the handsets have not such a long life as television sets, less than 3 years average compared with 10–15 years.

## C.2.1. *LTE terminals*

Smartphone sales have seen a remarkable growth over the past 3 years. Of course, this was mainly on 3G networks.

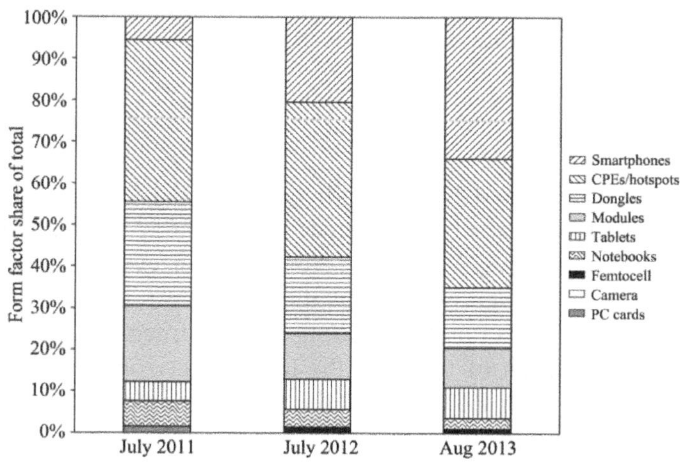


**Figure C.8.** *Devices*

On 4G, the market is today mainly in North America with Verizon Wireless and ATT Wireless leading the race. GSA has published a statistic evaluating the relative share of different devices as LTE terminals to date.

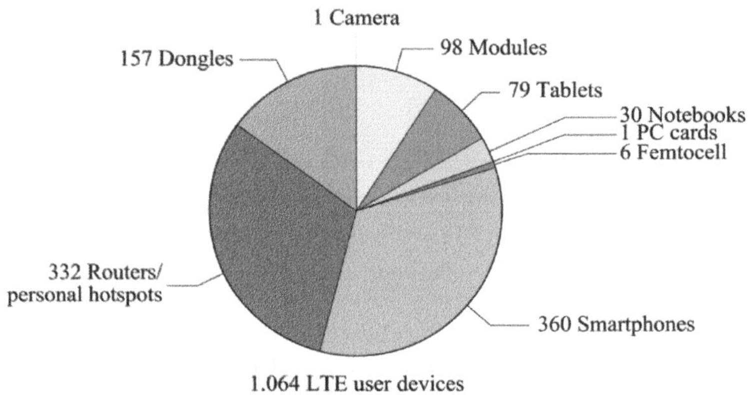


**Figure C.9.** *GSA analysis of the LTE terminal market end 2013*

This situation is not surprising since LTE was in release 8 without telephony services; so the actual networks were offering essentially a breakthrough in the mobile high-speed data transmission. It is quite normal that dongles and PC cards represent nearly half the number of LTE terminals with tablets, notebooks and routers making the share of computer-related devices reach more than 60%.

### C.2.2. *LTE coverage*

Verizon has opened the issue of using LTE for providing high-speed Internet in sparsely populated areas, where wiring would be dramatically expensive. With LTE Advanced, or even with LTE, the operator expects to serve farms in the Mid West at least 10 Mbps downlink and 5 Mbps uplink, avoiding an investment in kilometers of lines to connect each of them. In the towns, Verizon, of course, will cable all homes with FTTH since the length from the central office will be less than 1 km.

LTE in towns is supposed to be rolled out in higher frequencies for providing capacity. In a few cities (e.g. San Francisco), LTE will have to compete with Wi-Fi hot spots spread all over the dense area by the city council. This raises the question of mobility: to check a laptop or tablet when a person is sitting on a city train, there is no need for a system to be able to cope with 350 km/h speeding vehicles. LTE Advanced addresses this issue and considers that some low-speed services will be of interest.

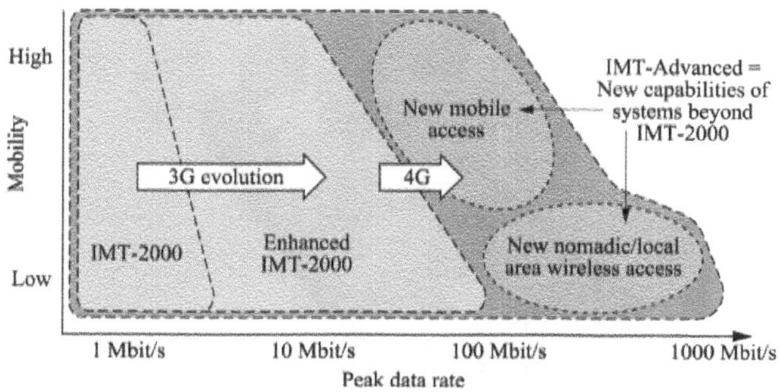


**Figure C.10.** *4G throughput goals as they apply to LTE were represented in the standard community by this picture nicknamed "the van" for its shape: it shows throughout evolution goals as a function of mobility speed*

In this figure, we must not that very high bit rate will not be provided to high-speed moves, for which the eNodeB will stay around 100 Mbps. However, the increased transmission rate will be looking at the market of nomadic users, e.g. airports, commercial malls and other gathering places for laptop addicts.

### C.2.3. *LTE and smartphones*

From the above survey, it can be concluded that the LTE market is split into two main domains:

– the computer-centric applications and devices;

– the general communication world, where the smartphone will be not only the king but also the only player. Maybe this trend is a little too manichean. Probably there will remain some irreducible people for whom a mobile handset's main function is telephony or SMS.

The smartphone will offer the kinds of service that are already available on 2G/3G networks.

**Figure C.11.** *Services offered on smartphones*

## C.3. Observed performances of LTE in already working networks

UMTS was also promising the Moon in 1998. Therefore, it seems interesting to check on what is happening on actually deployed and operated networks. The following are monitored throughputs observed on the four LTE networks, which are operated in France: Orange, SFR, Free and Bouygues Telecom (ByT). Measurements were carried out in December 2013.

(http://www.zdnet.fr/services/4g-monitor/)

*Top 10%*

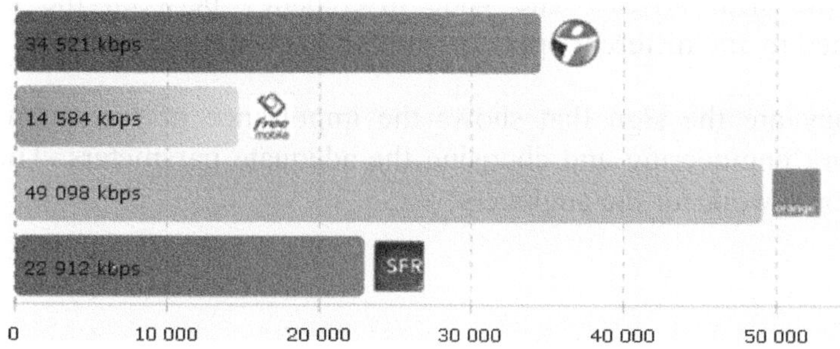


**Figure C.12.** *Mean downlink taken from the 10% faster test measured*

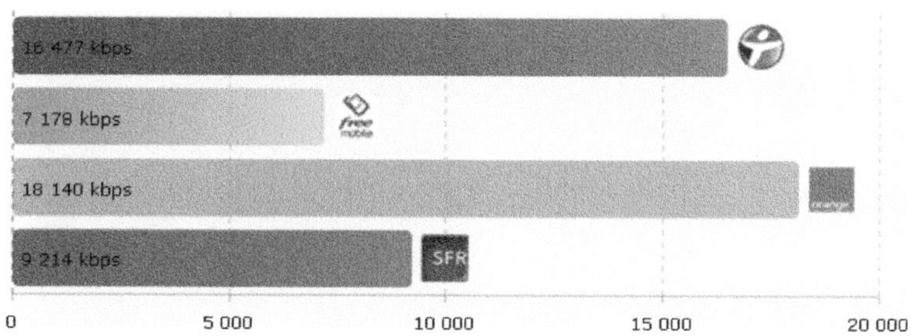


**Figure C.13.** *Mean downlink*

**Figure C.14.** *Mean uplink*

These measurements have to be considered carefully, but nevertheless they show that the subscriber's perception of LTE is not (yet) a service that compares with FTTH, but a better one than what is available with ADSL. The wide discrepancy between the results obtained by the different operators is also interesting.

They are the sign that shows the importance of fine-tuning the network engineering and choosing the adequate parameters. There is still some work for the engineers!

# Bibliography


[COR 97] Correia L.M., Prasad R., "An overview of wireless broadband communications", *IEEE Communications Magazine*, vol. 35, no. 1, pp. 28–33, January 1997.

[EKS 06] Ekstrom H., Furuskar A., Karlsson J., *et al.*, "Technical solutions for the 3G long-term evolution, Ericsson Res., Aachen, Germany", *IEEE Communications Magazine*, vol. 44, no. 3, pp. 38–45, March 2006.

[FCC 13] Federal Communications Commission, FCC adopts 15th report on video competition, Federal Communications Commission, 22 July 2013.

[GL 14] GL, MAPS$^{TM}$ Long Term Evolution (LTE) – (eGTP) Interfaces, www.gl.com/Brochures/Brochures/maps-lte-eGTP-brochure.pdf, 2014.

[KIR 83] Kirkpatrick S., Gelatt C.D., Vecchi M.P., "Optimization by simulated annealing", *Science*, vol. 220, no. 4598, pp. 671–680, 1983.

[KUM 98] Kumar S., Laha S., Nanda S., *et al.*, "Evolution of wireless data services: IS-95 to cdma2000", *IEEE Communications Magazine*, vol. 36, no. 10, pp. 140–149, October 1998.

[LUR 09] Luria R., *et al.*, "Cognitive effects of radiation emitted by cellular phones: the influence of exposure side and time", *Biolectromagnetics*, vol. 30, pp. 198–204, 2009.

[LUR 11] Luria R., *et al.*, "Effects of cell phone radiofrequency signal exposure on brain glucose metabolism", *Journal of the American Medical Association*, vol. 305, no. 8, 23 February 2011.

[RFC 00] RFC 2960, https://www.ietf.org/rfc/rfc2960.txt, 2000.

[3GP XX] 3GPP, http://www.3gpp.org/specifications.

# Index

Other titles from

ISTE

in

Networks and Telecommunications

## 2014

BITAM Salim, MELLOUK Abdelhamid
*Bio-inspired Routing Protocols for Vehicular Ad-Hoc Networks*

CAMPISTA Miguel Elias Mitre, RUBINSTEIN Marcelo Gonçalves
*Advanced Routing Protocols for Wireless Networks*

CHETTO Maryline
*Real-time Systems Scheduling 1: Fundamentals*
*Real-time Systems Scheduling 2: Focuses*

EXPOSITO Ernesto, DIOP Codé
*Smart SOA Platforms in Cloud Computing Architectures*

MELLOUK Abdelhamid, CUADRA-SANCHEZ Antonio
*Quality of Experience Engineering for Customer Added Value Services*

OTEAFY Sharief M.A., HASSANEIN Hossam S.
*Dynamic Wireless Sensor Networks*

REMY Jean-Gabriel, LETAMENDIA Charlotte
*LTE Standards*

TANWIR Savera, PERROS Harry
*VBR Video Traffic Models*

VAN METER Rodney
*Quantum Networking*

XIONG Kaiqi
*Resource Optimization and Security for Cloud Services*

## 2013

ASSING Dominique, CALÉ Stéphane
*Mobile Access Safety: Beyond BYOD*

BEN MAHMOUD Mohamed Slim, LARRIEU Nicolas, PIROVANO Alain
*Risk Propagation Assessment for Network Security: Application to Airport Communication Network Design*

BEYLOT André-Luc, LABIOD Houda
*Vehicular Networks: Models and Algorithms*

BRITO Gabriel M., VELLOSO Pedro Braconnot, MORAES Igor M.
*Information-Centric Networks: A New Paradigm for the Internet*

BERTIN Emmanuel, CRESPI Noël
*Architecture and Governance for Communication Services*

DEUFF Dominique, COSQUER Mathilde
*User-Centered Agile Method*

DUARTE Otto Carlos, PUJOLLE Guy
*Virtual Networks: Pluralistic Approach for the Next Generation of Internet*

FOWLER Scott A., MELLOUK Abdelhamid, YAMADA Naomi
*LTE-Advanced DRX Mechanism for Power Saving*

JOBERT Sébastien *et al.*
*Synchronous Ethernet and IEEE 1588 in Telecoms: Next Generation Synchronization Networks*

MELLOUK Abdelhamid, HOCEINI Said, TRAN Hai Anh
*Quality-of-Experience for Multimedia: Application to Content Delivery Network Architecture*

NAIT-SIDI-MOH Ahmed, BAKHOUYA Mohamed, GABER Jaafar, WACK Maxime
*Geopositioning and Mobility*

PEREZ André
*Voice over LTE: EPS and IMS Networks*

## 2012

AL AGHA Khaldoun
*Network Coding*

BOUCHET Olivier
*Wireless Optical Communications*

DECREUSEFOND Laurent, MOYAL Pascal
*Stochastic Modeling and Analysis of Telecoms Networks*

DUFOUR Jean-Yves
*Intelligent Video Surveillance Systems*

EXPOSITO Ernesto
*Advanced Transport Protocols: Designing the Next Generation*

JUMIRA Oswald, ZEADALLY Sherali
*Energy Efficiency in Wireless Networks*

KRIEF Francine
*Green Networking*

PEREZ André
*Mobile Networks Architecture*

## 2011

BONALD Thomas, FEUILLET Mathieu
*Network Performance Analysis*

CARBOU Romain, DIAZ Michel, EXPOSITO Ernesto, ROMAN Rodrigo
*Digital Home Networking*

CHABANNE Hervé, URIEN Pascal, SUSINI Jean-Ferdinand
*RFID and the Internet of Things*

GARDUNO David, DIAZ Michel
*Communicating Systems with UML 2: Modeling and Analysis of Network Protocols*

LAHEURTE Jean-Marc
*Compact Antennas for Wireless Communications and Terminals: Theory and Design*

RÉMY Jean-Gabriel, LETAMENDIA Charlotte
*Home Area Networks and IPTV*

PALICOT Jacques
*Radio Engineering: From Software Radio to Cognitive Radio*

PEREZ André
*IP, Ethernet and MPLS Networks: Resource and Fault Management*

TOUTAIN Laurent, MINABURO Ana
*Local Networks and the Internet: From Protocols to Interconnection*

## 2010

CHAOUCHI Hakima
*The Internet of Things*

FRIKHA Mounir
*Ad Hoc Networks: Routing, QoS and Optimization*

KRIEF Francine
*Communicating Embedded Systems / Network Applications*

## 2009

CHAOUCHI Hakima, MAKNAVICIUS Maryline
*Wireless and Mobile Network Security*

VIVIER Emmanuelle
*Radio Resources Management in WiMAX*

## 2008

CHADUC Jean-Marc, POGOREL Gérard
*The Radio Spectrum*

GAÏTI Dominique
*Autonomic Networks*

LABIOD Houda
*Wireless Ad Hoc and Sensor Networks*

LECOY Pierre
*Fiber-optic Communications*

MELLOUK Abdelhamid
*End-to-End Quality of Service Engineering in Next Generation Heterogeneous Networks*

PAGANI Pascal *et al.*
*Ultra-wideband Radio Propagation Channel*

## 2007

BENSLIMANE Abderrahim
*Multimedia Multicast on the Internet*

PUJOLLE Guy
*Management, Control and Evolution of IP Networks*

SANCHEZ Javier, THIOUNE Mamadou
*UMTS*

VIVIER Guillaume
*Reconfigurable Mobile Radio Systems*